Springer Theses

Recognizing Outstanding Ph.D. Research

Aims and Scope

The series "Springer Theses" brings together a selection of the very best Ph.D. theses from around the world and across the physical sciences. Nominated and endorsed by two recognized specialists, each published volume has been selected for its scientific excellence and the high impact of its contents for the pertinent field of research. For greater accessibility to non-specialists, the published versions include an extended introduction, as well as a foreword by the student's supervisor explaining the special relevance of the work for the field. As a whole, the series will provide a valuable resource both for newcomers to the research fields described, and for other scientists seeking detailed background information on special questions. Finally, it provides an accredited documentation of the valuable contributions made by today's younger generation of scientists.

Theses are accepted into the series by invited nomination only and must fulfill all of the following criteria

- They must be written in good English.
- The topic should fall within the confines of Chemistry, Physics, Earth Sciences, Engineering and related interdisciplinary fields such as Materials, Nanoscience, Chemical Engineering, Complex Systems and Biophysics.
- The work reported in the thesis must represent a significant scientific advance.
- If the thesis includes previously published material, permission to reproduce this must be gained from the respective copyright holder.
- They must have been examined and passed during the 12 months prior to nomination.
- Each thesis should include a foreword by the supervisor outlining the significance of its content.
- The theses should have a clearly defined structure including an introduction accessible to scientists not expert in that particular field.

More information about this series at http://www.springer.com/series/8790

Dinh Van Tuan

Charge and Spin Transport in Disordered Graphene-Based Materials

Doctoral Thesis accepted by
Autonomous University of Barcelona, Spain

 Springer

Author
Dr. Dinh Van Tuan
Catalan Institute of Nanoscience
  and Nanotechnology
Barcelona
Spain

Supervisor
Prof. Stephan Roche
Catalan Institute of Nanoscience
  and Nanotechnology
Barcelona
Spain

and

Case Western Reserve University
Cleveland, OH
USA

ISSN 2190-5053   ISSN 2190-5061   (electronic)
Springer Theses
ISBN 978-3-319-25569-9   ISBN 978-3-319-25571-2   (eBook)
DOI 10.1007/978-3-319-25571-2

Library of Congress Control Number: 2015952026

Springer Cham Heidelberg New York Dordrecht London

Printed on acid-free paper

Springer International Publishing AG Switzerland is part of Springer Science+Business Media
(www.springer.com)

I would like to dedicate this thesis to my loving parents

Supervisor's Foreword

Research into graphene started in 2005, following the discovery that a single monolayer of graphene could be separated out simply using a scotch tape-induced mechanical exfoliation of natural (but high quality) graphite raw material. The magnificent properties of graphene were first revealed by the inspiring and fecund theoretical analogy between hypothetical properties of relativistic particles (predicted but never observed, such as *the Klein tunnelling*) and the low-energy excitations in graphene, formally described as massless Dirac fermions (with a new quantum degree of freedom, namely the pseudospin). Many other unconventional and spectacular transport properties have been revealed during the past decade, including (to cite the most salient ones) weak antilocalization driven by the pseudospin degree of freedom, absence of strong localization and minimum conductivity, the half-integer quantum Hall effect, quantum Hall ferromagnetism, and the observation of the Hofstadter's butterfly for graphene deposited on boron nitride substrate. The simplicity (and beauty) of the graphene honeycomb lattice structure is in sharp contrast to the endless richness of fascinating transport phenomena, which today are still undergoing in-depth exploration.

On the other hand, the nature of disorder in such a peculiar two-dimensional material is manifold, and although charge transport in graphene is robust to a wide class of surrounding imperfections such as mechanical deformations, nearby charged impurities, or surface adsorbed (light) atoms, there exists another large category of defects which causes much more damage to the remarkable massless Dirac fermion transport physics. But beyond a simple transition to a bad conductor upon disordering, "more damaged" graphene can also exhibit unprecedented transport features especially when defects create low-energy impurity resonances which, given their real space long-range character, usually provoke new percolation paths for unconventional transport. This is particularly manifest in the high magnetic field regime, where exotic transport features appear in the phase diagram of the quantum Hall effect, or upon surface segregation of heavy ad-atoms (such as indium or thallium), which locally strongly enhances spin–orbit interaction, and by proximity effects that generate local energy bandgaps and frontier chiral states.

Additionally, quantifying the impact of disorder on charge mobilities and graphene device characteristics is essential for developing future graphene applications in flexible and transparent electronics, long-life batteries, or spintronics. Graphene spintronics actually stands as a particularly promising and challenging topic, as evidenced by the efforts of 2007 physics Nobel laureate Albert Fert, who is particularly active in the field, claiming that graphene is the perfect material for developing lateral spintronics and integrating novel types of non-charge-based information processing technologies, with hope for "*a second spintronic revolution*" after the discovery of giant magnetoresistance and its massive impact on storage technologies. This belief is shared by a pool of European researchers, who, together with Professor Fert, are engaged in the graphene spintronics workpackage of the Graphene Flagship EU-project, aiming to realize this visionary prediction (graphene-flagship.eu/).

Dinh Van Tuan has joined this adventure of the exploration of the fundamentals of charge and spin transport in disordered graphene in 2011, and has focused his Ph.D. on developing new theoretical methodologies to scrutinize the transport properties in large-scale models of disordered graphene, paying particular attention to fundamental defects such as grain boundaries in polycrystalline graphene, structural defects (often encountered in reduced graphene oxides), or various types of adsorbed ad-atoms (such as transition metal atoms or heavy atoms such as thallium). He has successfully, technically integrated spin–orbit interaction in the formalism of time-dependent evolution of wavepackets in real space, which has provided essential information about peculiarities of the spin dynamics of massless Dirac fermions.

The endeavor accomplished by Dinh, as evidenced by this high-quality thesis, has resulted in an outstanding piece of work, which has been acknowledged by many publications in high-impact journals, starting with the 2014 publication in Nature Physics of the discovery of an unprecedented mechanism for spin relaxation, unique to graphene and driven by spin/pseudospin entanglement (*D. Van Tuan et al., Nature Physics, 10, 857–863 (2014)*). This theoretical work shines new light on experimental controversies and has allowed us to revisit our understanding of spin transport phenomena in graphene-related materials, by demonstrating how weak spin–orbit interaction driven by transition metal ad-atoms could induce a complex spin and pseudospin dynamics at the origin of unique spin decoherence and relaxation effects.

The findings not only establish a more solid foundation for spin dynamics of massless Dirac fermions, but also open inspiring avenues towards the manipulation of the spin degree of freedom (by acting on pseudospin) and the realization of spin-based information processing technologies. A massive scientific impact of this work is thus expected, and the adventurous spirit and hard work of Dinh have been essential in this accomplishment. This thesis also presents several other high-level scientific results such as the impact of various types of defects on quantum transport in graphene, and particularly the effect of polycrystalline morphology on charge mobility (with the discovery of a new transport scaling law). Taken as a whole, it provides guidance and inspiration for current and future experimental work, with a likely collateral impact on the improvement of graphene device engineering and applications of graphene materials.

Barcelona
April 2015

Prof. Stephan Roche

Abstract

This thesis is focused on modeling and simulation of charge and spin transport in two-dimensional graphene-based materials as well as the impact of graphene polycrystallinity on the performance of graphene field-effect transistors. The Kubo–Greenwood transport approach has been used as the key method to carry out numerical calculations for charge transport properties. The study covers a wide range of disorders in graphene, from vacancies to chemical adsorbates on grain boundaries of polycrystalline graphene, and takes into account important quantum effects such as quantum interference and spin–orbit coupling effects. For spin transport, a new method based on the real space order $O(N)$ transport formalism is developed to explore the mechanism of spin relaxation in graphene. A new spin relaxation phenomenon related to spin/pseudospin entanglement is unveiled and could be the main mechanism at play governing fast spin relaxation in ultra-clean graphene.

Acknowledgments

First of all, I would like to express my deep gratitude to Prof. Dr. Stephan Roche for his assistance as mentor of my thesis. Without his kind encouragement, support, constructive guidance, and also proofreading the manuscript, I would not have been able to finish this thesis.

I would like to thank Dr. Frank Ortmann, Dr. David Soriano, Dr. Aron Cummings, Prof. Sergio Valenzuela, Prof. David Jiménez, Dr. Jani Kotakoski, Dr. Jose Eduardo Barrios Vargas, Dr. Nicolas Leconte, Prof. Pablo Ordejón, Dr. Alessandro Cresti, Prof. Jannik C. Meyer, Mr. Thibaud Louvet, Mr. Paweł Lenarczyk, Prof. Young Hee Lee, Dr. Dinh Loc Duong, Mr. Van Luan Nguyen, Dr. Ferney Chaves, Prof. M.F. Thorpe, and Dr. Avishek Kumar for their guidance, interesting discussions, suggestions, and collaborative work. I am also very thankful to Dr. Nicolas Leconte for taking his time to read and correct this thesis.

I will never forget the hospitality of Institut Català de Nanociència i Nanotecnologia (ICN2), and for that I would like to thank Mrs. Rosa Juan Nebot, Mrs. Anabel Rodrguez Sandá, Mrs. Inmaculada Caño Zafra, Mrs. Sandra Domene Megias, Mrs. Emma Nieto Fumanal, Mrs. Ana de la Osa Chaparro, and my dear colleagues in the Theoretical and Computational Nanoscience Group.

I acknowledge Prof. Jordi Pascual for accepting to be my tutor, Prof. David Jiménez, Prof. Francisco Paco Guinea, Prof. Jean-Christophe Charlier, Prof. Nicolas Lorente, Dr. Riccardo Rurali, Dr. Xavier Cartoixà Soler, and Dr. Xavier Waintal for accepting to be the jury members on my thesis defense.

Deep in my heart, I would like to thank my loving parents, my sisters, and my whole family for their love, support in all respects, and continuous encouragement, which is meaningful not only to my work but also to my life.

On the challenging road of my scientific life, I am happy and proud to have my wife, Nguyen Thi Thanh Thuy. She is always with me to share happiness as well as disappointment. I want to thank her for her understanding, support, and especially for her present of love, our daughter Dinh Quynh Chi.

Contents

Chapter 1
Introduction

Graphene, an atomic monolayer of carbon atoms arranged into a honeycomb lattice, is a fascinating and unique system. It is an extreme 2D condensed matter system where the charge carrier dynamics can be described as quasi-relativistic particles with zero effective carrier mass and the transport properties are governed by the Dirac equation, whereby their mobilities have unprecedentedly large values. Many of the interesting properties in graphene result from these characteristics which are analogous to those of relativistic, massless fermions. During the past ten years after its discovery, graphene has attracted a great attention. Ever since, numerous unique electrical, optical, and mechanical properties of graphene have been discovered such as optical transparence, high strength and stiffness, Klein tunneling, half-integer quantum Hall effect (QHE), weak antilocalization (WAL), etc. However, disorders are unavoidable factors that affect transport properties of graphene and it is crucial to study their detrimental effects to have a comprehension of real graphene samples.

Moreover, in order to develop technology and application based on graphene the integration of the material at wafer scale is mandatory. The chemical vapor deposition (CVD) growth technique is the best candidate for achieving a combination of high structural quality and wafer-scale growth. However, the resulting CVD graphene is polycrystalline graphene (Poly-G) [1–4], formed by many single-crystal grains with different orientations [5]. In order to accommodate the lattice mismatch between misoriented grains, the graphene grain boundaries (GGBs) in Poly-G are made up of a variety of non-hexagonal carbon rings, which can act as a source of scattering during charge transport. The properties of Poly-G are therefore dictated by their grain size and by the atomic structure at the grain boundaries (GBs). Effects of structural defects on the electronic, mechanical and transport properties of graphene have recently been analyzed theoretically [6, 7]. Moreover, several theoretical studies have reported on the effect of a single GB on electronic [8, 9], magnetic [10], chemical [11], and mechanical [12–14] properties of graphene. However, very few studies [11, 14] have discussed more complex forms of GBs (not restricted to infinite linear arrangements of dislocation cores), which would better correspond to the experimentally observed structures [5, 15, 16]. Furthermore, because of experimental challenges only a few experimental works [17] have systematically investigated the impact of GBs on

© Springer International Publishing Switzerland 2016
D.V. Tuan, *Charge and Spin Transport in Disordered Graphene-Based Materials*, Springer Theses, DOI 10.1007/978-3-319-25571-2_1

electronic transport, mainly confirming the reduced conductivity as compared to single-crystalline samples. Very recent electrical measurements on individual GBs in CVD-graphene also reported that a good interdomain connectivity is a fundamental geometrical requirement for improved transport capability [18]. However, to date little is known about the global contribution of complex distributions of GBs to measured charge mobilities [19]. Therefore, to understand the large-scale electrical transport properties of Poly-G, it is important to perform a detailed exploration of the role played by the GBs.

In regards to the potential of graphene for spintronics, the extremely small intrinsic spin-orbit coupling (SOC) of graphene and the lack of hyperfine interaction with the most abundant carbon isotope have led to intense research into possible applications of this material in spintronic devices with the anticipated possibility of transporting spin information over very long distances [20–22]. However, the spin relaxation times are still found to be orders of magnitude smaller than initially predicted [23–27], while the major physical process for spin equilibration and its dependence on charge density and disorder remain elusive. Experiments have been analyzed in terms of the conventional Elliot-Yafet (EY) and Dyakonov-Perel (DP) processes, yielding contradictory results. Recently, a mechanism based on resonant scattering by local magnetic moments has also been proposed [28] but contains too many free parameters and does not solve the controversial results reported experimentally [29].

In 2005, the quantum spin Hall (QSH) state was predicted in graphene by Kane and Mele [30]. The Kane and Mele model is two copies of the Haldane model [31] such that the spin up electron exhibits a chiral integer QHE while the spin down electron exhibits an anti-chiral integer QHE. This novel electronic state of matter is gapped in the bulk and supports the transport of spin and charge in gapless edge states that propagate at the sample boundaries. The edge states are insensitive to disorder (which does not break time reversal symmetry) because their directionality is correlated with spin. However, this beautiful state is unobservable in graphene due to weak SOC in intrinsic graphene. A solution for this problem is endowing graphene with certain heavy adatoms such as thallium or indium [32], but to date the clustering effect of these adatoms make the QSH state seem to jeopardize its observation.

The purpose of this thesis is to address above problems. The thesis is organized into 6 chapters and 2 appendices. The contents are developed as follows:

This chapter gives the purpose of this thesis and overviews the problems of interest. The content of each chapter in this thesis is also mentioned in this introductory chapter.

Chapter 2 presents the electronic and transport properties of clean graphene. In this chapter the linear band structure of graphene is derived, and some special characteristics of Dirac fermions such as chirality, zero effective mass, etc. are mentioned. The chapter also covers the literature of electronic transport and spin transport in graphene. In this later part, spin-orbit interactions are derived and their modifications on the Dirac band structure are reviewed. The final part of this chapter is devoted to a discussion on the discrepancy of experimental and theoretical results concerning spin relaxation in graphene. Two mechanisms for spin relaxation in graphene, EY and DP, are also derived.

Chapter 3 briefly overviews the Kubo-Greenwood transport formalism which is extensively used in this thesis. In this chapter, two different approaches are discussed namely, the semiclassical and quantum approaches, which lead to the Einstein relation for conductivity. The real space transport method for the Kubo conductivity calculation is also introduced. An extension of real space order $O(N)$ transport formalism is developed to study spin transport in the realistic system.

Chapter 4 focuses on the electronic transport properties of disordered graphene. The transport properties are studied with gradually increasing disorder, from point defects in graphene with vacancies to line defects in Poly-G and finally to the extremely disordered form of graphene, amorphous membranes of sp^2 graphene. The studies are systematically concentrated on different aspects of graphene in perspectives of applications.

Chapter 5 deals with the graphene spin relaxation problems. In this chapter we point out the limitations of EY and DP mechanisms for graphene, and we propose a new mechanism driven by the entanglement between spin and pseudospin quantum degree of freedoms, which governs the fast spin relaxation close to Dirac point in graphene. At the end of this chapter, we explain the difficulty of observing the QSH effect in graphene when depositing heavy adatoms. The natural clustering trend of such adatoms weakens the SOC effect which is a crucial factor of the formation of topological edge state. The chapter also reports the formation of a robust metallic state which is related to the enhanced percolation of propagating states between islands.

Chapter 6 summarizes the thesis and suggests some opening directions for the near future.

References

1. X.S. Li, W.W. Cai, J.H. An, S. Kim, J. Nah, D.X. Yang, R. Piner, A. Velamakanni, I. Jung, E. Tutuc, S.K. Banerjee, L. Colombo, R.S. Ruoff, Science **324**, 1312–1314 (2009)
2. A. Reina, X. Jia, J. Ho, D. Nezich, H. Son, V. Bulovic, M.S. Dresselhaus, J. Kong, Nano Lett. **9**, 30 (2009)
3. X.S. Li, C.W. Magnuson, A. Venugopal, J.H. An, J.W. Suk, B.Y. Han, M. Borysiak, W.W. Cai, A. Velamakanni, Y.W. Zhu, L.F. Fu, E.M. Vogel, E. Voelkl, L. Colombo, R.S. Ruoff, Nano Lett. **10**, 4328–4334 (2010)
4. S. Bae, H. Kim, Y. Lee, X.F. Xu, J.S. Park, Y. Zheng, J. Balakrishnan, T. Lei, H.R. Kim, Y.I. Song, Y.J. Kim, K.S. Kim, B. Ozyilmaz, J.H. Ahn, B.H. Hong, S. Iijima, Nat. Nanotechnol. **5**, 574–578 (2010)
5. P.Y. Huang, C.S. Ruiz-Vargas, A.M. van der Zande, W.S. Whitney, M.P. Levendorf, J.W. Kevek, S. Garg, J.S. Alden, C.J. Hustedt, Y. Zhu, J. Park, P.L. McEuen, D.A. Muller, Nature **469**, 389 (2011)
6. A.V. Krasheninnikov, F. Banhart, Nat. Mater. **6**, 723 (2007)
7. S. Roche, N. Leconte, F. Ortmann, A. Lherbier, D. Soriano, J.-C. Charlier, Solid State Commun. **152**, 1404–1410 (2012)
8. N.M.R. Peres, F. Guinea, A.H.C. Neto, Phys. Rev. B **73**, 125411 (2006)
9. O.V. Yazyev, S.G. Louie, Nat. Mater. **9**, 806 (2010)
10. J. Cervenka, M.I. Katsnelson, C.F.J. Flipse, Nat. Phys. **5**, 840 (2009)

11. S. Malola, H. Hakkinen, P. Koskinen, Phys. Rev. B **81**, 165447 (2010)
12. Y. Liu, B.I. Yakobson, Nano Lett. **10**, 2178 (2010)
13. R. Grantab, V.B. Shenoy, R.S. Ruoff, Science **330**, 946 (2010)
14. J. Kotakoski, J.C. Meyer, Phys. Rev. B **85**, 195447 (2012)
15. K. Kim, Z. Lee, W. Regan, C. Kisielowski, M.F. Crommie, A. Zettl, ACS Nano **5**, 2142 (2011)
16. S. Kurasch, J. Kotakoski, O. Lehtinen, V. Skakalova, J. Smet, C.E. Krill, A.V. Krasheninnikov, U. Kaiser, Nano Lett. **12**, 3168 (2012)
17. Q. Yu et al., Nat. Mater. **10**, 443–449 (2011)
18. A.W. Tsen et al., Science **336**, 1143–1146 (2012)
19. A. Ferreira, X. Xu, C.-L. Tan, S. Bae, N.M.R. Peres, B.-H. Hong, B. Ozyilmaz, A.H. Castro Neto, Eur. Phys. Lett. **94**, 28003 (2011)
20. H. Min et al., Phys. Rev. B **74**, 165310 (2006)
21. C. Ertler et al., Phys. Rev. B **80**, 041405 (2009)
22. A.H. Castro Neto, F. Guinea, Phys. Rev. Lett. **103**, 026804 (2009)
23. N. Tombros et al., Nature (London) **448**, 571 (2007)
24. A. Avsar et al., Nano Lett. **11**, 2363–2368 (2011)
25. W. Han, R.K. Kawakami, Phys. Rev. Lett. **107**, 047207 (2011)
26. P.J. Zomer, M.H.D. Guimaraes, N. Tombros, B.J. van Wees, Phys. Rev. B **86**, 161416(R) (2012)
27. B. Dlubak, M.-B. Martin, C. Deranlot, B. Servet, S. Xavier, R. Mattana, M. Sprinkle, C. Berger, W.A. De Heer, F. Petroff, A. Anane, P. Seneor, A. Fert, Nat. Phys. **8**, 557 (2012)
28. D. Kochan, M. Gmitra, J. Fabian, Phys. Rev. Lett. **112**, 116602 (2014)
29. M. Wojtaszek, I.J. Vera-Marun, T. Maassen, B.J. van Wees, Phys. Rev. B **87**, 081402(R) (2013)
30. C.L. Kane, E.J. Mele, Phys. Rev. Lett. **95**, 226801 (2005)
31. F.D.M. Haldane, Phys. Rev. Lett. **61**, 2015 (1988)
32. C. Weeks, J. Hu, J. Alicea, M. Franz, R. Wu, Phys. Rev. X **1**, 021001 (2011)

Chapter 2
Electronic and Transport Properties of Graphene

2.1 Introduction

Graphene has received a great attention since it was first isolated by Nobel Laureates Konstantin Novoselov and Andre K. Geim in 2004. The reason for such excitement is that graphene is the first truly 2D crystal ever observed in nature and possesses remarkable electrical, chemical and mechanical properties. Furthermore, electrons in graphene show a quasi-relativistic behavior, and the system is therefore an ideal candidate for the test of quantum field-theoretical models that have been developed in high-energy physics. Most prominently, electrons in graphene may be viewed as massless charged fermions existing in 2D space, particles that one usually does not encounter in our three-dimensional world. Indeed, all massless elementary particles, such as photons or neutrinos, happen to be electrically neutral. Graphene is therefore an exciting bridge between condensed matter and high-energy physics, and the research on its electronic properties unites scientists with various thematic backgrounds.

Graphene is also an attractive material for spintronics due to the theoretical possibility of long spin lifetimes arising from low intrinsic SOC and weak hyperfine interaction [1]. However, Hanle spin precession measurements and non-local spin valve geometry have reported spin lifetimes that are orders of magnitude shorter than expected theoretically [2–5]. Several studies have investigated spin relaxation including the roles of impurity scattering [5] and graphene thickness [6] and specially, ferromagnet contact-induced spin relaxation was predicted to be responsible for the short spin lifetimes observed in experiments [7]. However, these explanations have not given a satisfying answer for the discrepancy between theoretical results and experimental data. This has prompted theoretical studies of the extrinsic sources of spin relaxation such as impurity scattering [8], ripples [1], and substrate effects [9]. The problem remains however still puzzling and unsolved.

In this chapter, we will briefly review some theoretical and experimental results about fundamental electric and spin transport properties of graphene. Firstly, we will derive graphene band structure and massless Dirac equation for graphene in

D.V. Tuan, *Charge and Spin Transport in Disordered Graphene-Based Materials*, Springer Theses, DOI 10.1007/978-3-319-25571-2_2

Sect. 2.2. Next, some experimental and theoretical studies about transport properties of graphene are discussed in Sect. 2.3. Section 2.4 discusses some aspects of SOC in graphene, which plays an important role for studying spin relaxation in Chap. 5.

2.2 Graphene and Dirac Fermions

The most interesting property of graphene might be the Dirac-cone energy dispersion. This is the consequence of sp^2 hybridization and graphene symmetry. In this section, I briefly review its structure, the commonly used tight-binding (TB) description and the deviation of the linear energy dispersion of graphene.

2.2.1 Graphene

Graphene is a single atomic layer of graphite, an allotrope of carbon that is made up of very tightly bonded carbon atoms organised into a hexagonal lattice. What makes graphene so special is its sp^2 hybridization and very thin atomic thickness (see Fig. 2.1). These properties are what enable graphene to break so many records in terms of strength, electricity, heat conduction, etc.

Carbon is a common element in the nature, with atomic number 6, group 14 on the periodic table. The electronic configuration of carbon is $1s^2 2s^2 2p^2$ which shows that carbon has 4 electrons ($2s$ and $2p$) in its outer shell which is available for forming chemical bonds. In graphene, these four valence electrons form sp^2 hybridization in which three electrons are distributed into three in-plane σ bonds, which are strongly covalent, determining the energetic stability and the elastic properties of graphene.

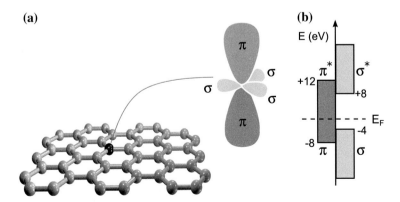

Fig. 2.1 Electronic structure of graphene **a** Graphene sample and the sp^2 hybridization in graphene. **b** Energy range of orbitals in graphene (Figure is taken from [10])

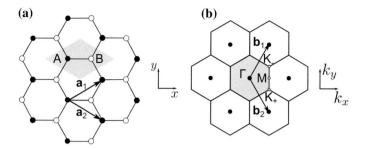

Fig. 2.2 Real (**a**) and reciprocal (**b**) space of graphene lattice (Figure is taken from [10])

The remaining electron in the p_z orbitals, which is perpendicular to graphene plane, forms the π bond in graphene (See Fig. 2.1).

The calculation for the energy ranges of σ and π bands (See Fig. 2.1b) shows that only electrons in the π bond contribute to the electronic properties of graphene because the σ bands are far away from the Fermi level. Because of this, it is sufficient to treat graphene as a collection of atoms with single p_z orbitals per site.

In graphene, carbon atoms are located at the vertices of a hexagonal lattice. Graphene is a bipartite lattice which consists of two sublattices A and B and basis vectors $(\mathbf{a}_1, \mathbf{a}_2)$ (See Fig. 2.2):

$$\mathbf{a}_1 = a\left(\frac{\sqrt{3}}{2}, \frac{1}{2}\right), \mathbf{a}_2 = a\left(\frac{\sqrt{3}}{2}, -\frac{1}{2}\right), \tag{2.1}$$

with $a = \sqrt{3}a_{cc}$, where $a_{cc} = 1.42$ Å is the carbon-carbon distance in graphene. These basis vectors build a hexagonal Brillouin zone with two inequivalent points K and K' (K_+ and K_- respectively in Fig. 2.2) at the corners

$$\mathbf{K} = \frac{4\pi}{3a}\left(\frac{\sqrt{3}}{2}, -\frac{1}{2}\right), \mathbf{K}' = \frac{4\pi}{3a}\left(\frac{\sqrt{3}}{2}, \frac{1}{2}\right), \tag{2.2}$$

As mentioned above and from Bloch's theorem, we can write the wave function in the form of p_z orbitals wave function at sublattices A $(\varphi(\mathbf{r} - \mathbf{r}_A))$ and B $(\varphi(\mathbf{r} - \mathbf{r}_B))$

$$\Psi(\mathbf{k}, \mathbf{r}) = c_A(\mathbf{k})\phi^A(\mathbf{k}, \mathbf{r}) + c_B(\mathbf{k})\phi^B(\mathbf{k}, \mathbf{r}) \tag{2.3}$$

where

$$\phi^A(\mathbf{k}, \mathbf{r}) = \frac{1}{\sqrt{N}}\sum_{\mathbf{R}_j} e^{i\mathbf{k}\cdot\mathbf{R}_j}\varphi(\mathbf{r} - \mathbf{r}_A - \mathbf{R}_j), \tag{2.4}$$

$$\phi^B(\mathbf{k}, \mathbf{r}) = \frac{1}{\sqrt{N}}\sum_{\mathbf{R}_j} e^{i\mathbf{k}\cdot\mathbf{R}_j}\varphi(\mathbf{r} - \mathbf{r}_B - \mathbf{R}_j), \tag{2.5}$$

where $\mathbf{k}$ is the electron wavevector, N the number of unit cells in the graphene sheet, and $\mathbf{R}_j$ is a Bravais lattice point.

Using the Schrödinger equation, $\mathcal{H}\Psi(\mathbf{k}, \mathbf{r}) = E\Psi(\mathbf{k}, \mathbf{r})$, one obtains a 2×2 eigenvalue problem,

$$\mathcal{H}(\mathbf{k})\begin{pmatrix} c_A(\mathbf{k}) \\ c_B(\mathbf{k}) \end{pmatrix} = \begin{pmatrix} \mathcal{H}_{AA}(\mathbf{k}) & \mathcal{H}_{AB}(\mathbf{k}) \\ \mathcal{H}_{BA}(\mathbf{k}) & \mathcal{H}_{BB}(\mathbf{k}) \end{pmatrix} \begin{pmatrix} c_A(\mathbf{k}) \\ c_B(\mathbf{k}) \end{pmatrix} = E(\mathbf{k}) \begin{pmatrix} S_{AA}(\mathbf{k}) & S_{AB}(\mathbf{k}) \\ S_{BA}(\mathbf{k}) & S_{BB}(\mathbf{k}) \end{pmatrix} \begin{pmatrix} c_A(\mathbf{k}) \\ c_B(\mathbf{k}) \end{pmatrix}.$$
(2.6)

where $S_{\alpha\beta}(\mathbf{k}) = \langle \phi^\alpha(\mathbf{k})|\phi^\beta(\mathbf{k})\rangle$ and the matrix elements of the Hamiltonian are given by:

$$\mathcal{H}_{AA}(\mathbf{k}) = \frac{1}{N} \sum_{\mathbf{R}_i, \mathbf{R}_j} e^{i\mathbf{k}.(\mathbf{R}_j - \mathbf{R}_i)} \langle \varphi^{A,\mathbf{R}_i} \mid \mathcal{H} \mid \varphi^{A,\mathbf{R}_j} \rangle \tag{2.7}$$

$$\mathcal{H}_{AB}(\mathbf{k}) = \frac{1}{N} \sum_{\mathbf{R}_i, \mathbf{R}_j} e^{i\mathbf{k}.(\mathbf{R}_j - \mathbf{R}_i)} \langle \varphi^{A,\mathbf{R}_i} \mid \mathcal{H} \mid \varphi^{B,\mathbf{R}_j} \rangle, \tag{2.8}$$

with $\mathcal{H}_{AA} = \mathcal{H}_{BB}$ and $\mathcal{H}_{AB} = \mathcal{H}_{BA}^*$, and introducing the notation: $\varphi^{A,\mathbf{R}_i} = \varphi(\mathbf{r} - \mathbf{r}_A - \mathbf{R}_i)$ and $\varphi^{B,\mathbf{R}_i} = \varphi(\mathbf{r} - \mathbf{r}_B - \mathbf{R}_i)$.

If we neglect the overlap $s = \langle \varphi^A | \varphi^B \rangle$ between neighboring p_z orbitals, then, $S_{\alpha\beta}(\mathbf{k}) = \delta_{\alpha,\beta}$ and Eq. (2.6) becomes

$$\begin{pmatrix} \mathcal{H}_{AA}(\mathbf{k}) & \mathcal{H}_{AB}(\mathbf{k}) \\ \mathcal{H}_{BA}(\mathbf{k}) & \mathcal{H}_{BB}(\mathbf{k}) \end{pmatrix} \begin{pmatrix} c_A(\mathbf{k}) \\ c_B(\mathbf{k}) \end{pmatrix} = E(\mathbf{k}) \begin{pmatrix} c_A(\mathbf{k}) \\ c_B(\mathbf{k}) \end{pmatrix}. \tag{2.9}$$

If we consider only the first-nearest-neighbors interactions then

$$\mathcal{H}_{AB}(\mathbf{k}) = \langle \varphi^{A,0}|\mathcal{H}|\varphi^{B,0} \rangle + e^{-i\mathbf{k}.\mathbf{a}_1} \langle \varphi^{A,0}|\mathcal{H}|\varphi^{B,-\mathbf{a}_1} \rangle + e^{-i\mathbf{k}.\mathbf{a}_2} \langle \varphi^{A,0}|\mathcal{H}|\varphi^{B,-\mathbf{a}_2} \rangle$$
$$= -\gamma_0\alpha(\mathbf{k}) \tag{2.10}$$

where γ_0 stands for the transfer integral between first neighbors π orbitals ($\gamma_0 = 2.7eV$ in this thesis) and $\alpha(\mathbf{k})$ is given by:

$$\alpha(\mathbf{k}) = (1 + e^{-i\mathbf{k}.\mathbf{a}_1} + e^{-i\mathbf{k}.\mathbf{a}_2}). \tag{2.11}$$

Taking $\mathcal{H}_{AA}(\mathbf{k}) = \mathcal{H}_{BB}(\mathbf{k}) = 0$ as the energy reference, we can write $\mathcal{H}(\mathbf{k})$ as:

$$\mathcal{H}(\mathbf{k}) = \begin{pmatrix} 0 & -\gamma_0\alpha(\mathbf{k}) \\ -\gamma_0\alpha(\mathbf{k})^* & 0 \end{pmatrix}. \tag{2.12}$$

Diagonalizing this Hamiltonian gives the energy dispersion relations for π^* (conduction) band (+) and π (valence) band (−):

$$E^\pm(\mathbf{k}) = \pm\gamma_0|\alpha(\mathbf{k})|$$
$$= \pm\gamma_0\sqrt{3 + 2\cos(\mathbf{k}.\mathbf{a}_1) + 2\cos(\mathbf{k}.\mathbf{a}_2) + 2\cos(\mathbf{k}.(\mathbf{a}_2 - \mathbf{a}_1))}$$
$$= \pm\gamma_0\sqrt{1 + 4\cos\frac{\sqrt{3}k_x a}{2}\cos\frac{k_y a}{2}4\cos^2\frac{k_y a}{2}}. \tag{2.13}$$

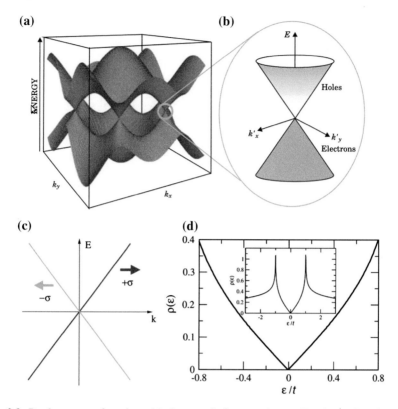

Fig. 2.3 Band structure of graphene (**a**), the zoom-in figure at close to K and K' points (**b**, **c**) and the density of state of graphene (**d**) (Figure is taken from [11])

This band structure is plotted in Fig. 2.3 with the symmetry between the conduction band and the valence band which touch at three K and K' points with zero density of state at this energy (Fig. 2.3d). Because of this, graphene is called gapless semiconductor or semi-metal. In neutral graphene, the Fermi level lie exactly at these points.

2.2.2 Low-Energy Dispersion

Because of the fact that they can only experimentally tune the Fermi level a small range (0.3 eV) about the touching points, this corresponds to a small variation around the K and K' points in momentum space. Therefore, it is sufficient to expand the energy dispersion in the vicinity of K and K' points by replacing $\mathbf{k} \rightarrow \mathbf{K}(\mathbf{K}') + \mathbf{k}$, which lets us write Eq. (2.12) in the form

$$\mathcal{H} = \hbar v_F (\eta \sigma_x k_x + \sigma_y k_y). \tag{2.14}$$

and Eq. (2.13) becomes

$$E_s(\mathbf{k}) = s\hbar v_F |\mathbf{k}|, \tag{2.15}$$

where $v_F = \sqrt{3}\gamma_0 a/2\hbar$ is the electronic group velocity, $\eta = 1(-1)$ for $K(K')$ points, $s = \pm 1$ is the band index (+1 for conduction band and -1 for valence band) and the Pauli matrices are defined as usual:

$$\sigma_x = \begin{pmatrix} 0 & 1 \\ 1 & 0 \end{pmatrix}, \quad \sigma_y = \begin{pmatrix} 0 & -i \\ i & 0 \end{pmatrix}, \quad \sigma_z = \begin{pmatrix} 1 & 0 \\ 0 & -1 \end{pmatrix}. \tag{2.16}$$

Equation (2.14) is almost the same the Dirac equation for the massless fermions in quantum electrodynamics except for the fact that the Pauli matrices here represent the sublattice degrees of freedom instead of spin and the speed of light c is replaced by graphene velocity $v_F \simeq c/300$. Therefore, the sublattice degrees of freedom and the touching points are called *pseudospin* and *Dirac point*, respectively.

The linear energy dispersion in Eq. (2.15) leads to the fact that total density of states is directly proportional to energy and carrier density is proportional to energy squared.

Indeed,

$$\rho(E) = \frac{1}{L^2} \sum_{\mathbf{k}} \delta(E - E(\mathbf{k})) = \int g_s g_v \frac{2\pi k dk}{(2\pi)^2} \delta(E - E(k)) = \frac{2|E|}{\pi \hbar^2 v_F^2} \tag{2.17}$$

which is plotted in Fig. 2.3d, where $g_s = 2$ and $g_v = 2$ account for spin and valley degeneracies, respectively. The carrier density is given by

$$n(E) = \frac{1}{L^2} \sum_{|\mathbf{k}| \leq k_F} g_s g_v = g_s g_v \frac{k_F^2}{4\pi} = \frac{E^2}{\pi \hbar^2 v_F^2} \tag{2.18}$$

To find the eigenstates of Dirac Hamiltonian (2.14), it is useful to write this Hamiltonian in the term of momentum direction θ_k

$$\mathcal{H}_\eta(\mathbf{k}) = \hbar v_F k \begin{pmatrix} 0 & e^{-i\eta\theta_k} \\ e^{+i\eta\theta_k} & 0 \end{pmatrix} \tag{2.19}$$

where $\theta_k = arctan(k_y/k_x)$. This Hamiltonian has the eigenvalues given by Eq. (2.15) and the eigenfunctions

$$|\Psi_{\eta,s}(\mathbf{k})\rangle = \frac{1}{\sqrt{2}} \begin{pmatrix} 1 \\ se^{i\eta\theta_k} \end{pmatrix}. \tag{2.20}$$

Next, we are going to find eigenvalues of the helicity operator (a very important feature of Dirac particle) which here is defined as:

$$\hat{h} = \hat{\sigma} \cdot \frac{\mathbf{p}}{|\mathbf{p}|}. \tag{2.21}$$

where $\mathbf{p} = \hbar\mathbf{k}$ is the electron momentum operator.

In order to do that, it is convenient to exchange the spinor components at the K' point (for $\eta = -1$) [12],

$$|\Psi^{\mathbf{K}}(\mathbf{k})\rangle = \begin{pmatrix} c_A(\mathbf{k}) \\ c_B(\mathbf{k}) \end{pmatrix}, \quad |\Psi^{\mathbf{K}'}(\mathbf{k})\rangle = \begin{pmatrix} c_B(\mathbf{k}) \\ c_A(\mathbf{k}) \end{pmatrix} \tag{2.22}$$

i.e., to invert the role of the two sublattices. In this case, the effective low-energy Hamiltonian in Eq. (2.14) may be represented as

$$\mathcal{H}^\eta(\mathbf{k}) = \eta\hbar v_F(\sigma_x k_x + \sigma_y k_y) = \hbar v_F \tau^z \otimes \vec{\sigma}\vec{k}. \tag{2.23}$$

where τ are Pauli matrices representing the valley degree of freedoms called *valley pseudospin*. Using Eqs. (2.23) and (2.21)

$$\mathcal{H}^\eta(\mathbf{k}) = \eta\hbar v_F k\hat{h} \tag{2.24}$$

we find that helicity operator commutes with the Hamiltonian, the projection of the pseudospin is a well-defined conserved quantity which can be either positive or negative, corresponding to pseudospin and momentum being *parallel* or *antiparallel* to each other. The band index s, which describes the valence and conduction bands, is therefore entirely determined by the chirality and the valley pseudospin, and one finds

$$s = \eta h \tag{2.25}$$

which help us find out that chirality changes sign from conduction band to valence band and from K to K' points. The fact that pseudospin is blocked with momentum has a strong influence in many of the most intriguing properties of graphene. For example, for an electron to backscatter (i.e. changing $\mathbf{p}$ to $-\mathbf{p}$) it needs to reverse its pseudospin (see Fig. 2.3c). So backscattering is not possible if the Hamiltonian is not perturbed by a term which flips the pseudospin. This makes electrons in graphene insensitive to long-range scatterers. This characteristic manifests itself in some phenomena such as Klein tunneling or WAL [13, 14]. Klein tunneling [15] is a spectacular manifestation of the Dirac fermions physics which describes that when the Dirac charge crosses a tunneling barrier, the incoming electron is partially or totally transmitted depending on the incident angle of the incoming wavepacket. Especially, the barrier always remains perfectly transparent for angles close to normal incidence regardless of the height and width of the barrier, standing as a feature unique to massless Dirac fermions and being completely different form the usual charge whose transmission probability decays exponentially with the barrier width. Klein tunneling has been studied theoretically and it shows that for long range potentials which preserve AB symmetry and prohibits intervalley scattering, the backscattering is totally suppressed.

In next section, we will discuss in more detail the effect of special band structure and pseudospin-momentum coupling on the transport properties of graphene.

2.3 Electronic and Transport Properties of Disordered Graphene

The disorder in graphene sample is practically an inevitable factor in any experiment. In some ways, artificial disorders are also tools to engineer, functionalize the materials. For instance, pure semiconductors are poor conductors and poor insulators. However, their magnificent properties have been achieved by functionalization using $n-$ and $p-$type dopants, leading to $p-n$ junctions, transistors, junction lasers, light-emitting diodes, and an entire technological revolution.

Similarly to semiconductors, in spite of having unique properties such as superb mechanical strength and carrier mobility, pristine graphene is not useful for practical applications because of its low carrier density, zero band gap, and chemical inertness. The lack of electronic gap in pristine graphene is an issue that has to be overcome to achieve high I_{on}/I_{off} current ratio in graphene-based field-effect devices. Therefore, it is important to study the disorder effect on the electronic properties of graphene not only to conquer its detrimental effects but also use artificial defects to functionalize graphene devices.

The study of transport properties is at the heart of graphene research. Experiments show that the conductivity (down to a few Kelvin) is almost constant close to the Dirac point, $\sigma \sim 2-5e^2/h$, and weakly dependent on the value of the charge mobility [15–18]. On the theoretical side, within the self-consistent Born approximation (SCBA) the semiclassical part of the conductivity due to short range disorder is found to be $\sigma^{min} = 4e^2/\pi h$ which is known as the quantum limited conductivity of graphene in clean limit [19].

However, transport properties of graphene are strongly dependent on the nature of possible sources of disorder. There are many kinds of disorders in graphene, some are long-range disorders such as Coulomb interactions of charged impurities in the substrate, electron-hole puddle, long range strain deformations, distortion of graphene structure, etc. Other forms are related to the sp^3 defects such as epoxide defects, the absorption of hydroxyl, hydrogen, fluorine, etc. on graphene (See Fig. 2.4). Finally topological disorders which keep the sp^2 hybridization of graphene but change the hexagonal structure, involve structural point defects and line defects or GBs.

The electronic properties of graphene are well described by the $\pi-$orbital tight-binding Hamiltonian in which the disorder in the real sample can be simulated by changing the on-site energies $\delta\epsilon$ of $\pi-$orbital. One of the simplest disorder model in

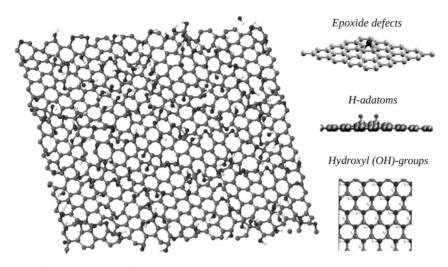

Fig. 2.4 Some kinds of sp^3 disorder in graphene

Fig. 2.5 Main frame: The semiclassical conductivity for Anderson disorder. *Inset* the comparison of Kubo-Greenwood approach with the Boltzmann and self-consistent Born approximation (Figure is taken from [20])

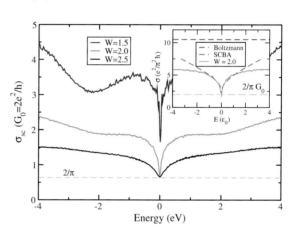

graphene is the short-range scattering potential, namely the Anderson disorder [20, 21]. This white noise uncorrelated disorder is introduced through random modulations of the onsite energies $\delta\epsilon \in [-W/2, W/2]\gamma_0$. This disorder could in principle mimic neutral impurities such as structural defects, dislocation lines, or adatoms, although the local geometry and chemical reactivity of defects and impurities actually demand for more sophisticated ab initio calculations if aiming at quantitative predictions.

Figure 2.5 shows the energy dependence of semiclassical conductivity from the Kubo formalism for some values of W (main frame) and the comparison with one from SCBA (inset). The results are in good agreement at low energy and close to the theoretically predicted minimum conductivity $\sigma^{min} = 4e^2/\pi h$. For higher energies, the agreement with SCBA is lost due to higher order deviations. Furthermore, the SCBA is not sufficient to describe such a system with all symmetries broken.

The role of disorder on WL and WAL in graphene has also been intensively investigated. From a general perspective, the conductance of a system can be viewed as the sum $(\mathcal{P}_{A \to B})$ over all probability amplitudes of propagating trajectories starting from one location A and going to another one B in real space, or more explicitly

$$G = \frac{2e^2}{h} \mathcal{P}_{A \to B} \tag{2.26}$$

$$\mathcal{P}_{A \to B} = \sum_i |\mathcal{A}_i|^2 + \sum_{i \neq j} \mathcal{A}_i \mathcal{A}_j^* \tag{2.27}$$

Here $\mathcal{A}_i = |\mathcal{A}_i| e^{i \Delta \varphi_i}$ is the propagation amplitude along the path i. The first term denotes the classical probability corresponding to semiclassical conductivity σ_{sc} while the second one is the interference term which gives the quantum correction $\delta\sigma(L)$ of the semiclassical result $\sigma(L) = \sigma_{sc} + \delta\sigma(L)$. For the majority of the trajectories the phase gains,

$$\Delta \varphi = \hbar^{-1} \int_A^B \mathbf{p} d\mathbf{l} \gg 1 \tag{2.28}$$

and the interference term vanishes. However, for some special trajectories with self-crossings, if we change the direction of propagation, $\mathbf{p} \to -\mathbf{p}$, $d\mathbf{l} \to -d\mathbf{l}$, the phase gains are the same, i.e. $\mathcal{A}_i \mathcal{A}_j^* = |\mathcal{A}_i|^2$, and quantum interference thus eventually enhance the probability of return to some origin. This contribution of quantum interferences gives rise to the increase of the quantum resistance, i.e. $\delta\sigma(L) < 0$, known as localization. There are two different scalling behaviors of localization: the WL with [10, 20, 22]

$$\delta\sigma(L) = -\frac{2e^2}{\pi h} \ln \left(\frac{L}{l_e} \right) \tag{2.29}$$

and strong localization described by

$$\sigma(L) \sim \exp \left(-\frac{L}{\xi} \right) \tag{2.30}$$

where L, l_e, and ξ denote the sample length, mean free path and the localization length, respectively.

However, we haven't considered the contribution of additional degrees of freedom such as spin or pseudospin in graphene. The detail calculations (for more details, see Ref. [10] and references therein) showed that these contributions can lead to the sign reversal of quantum correction of conductivity

$$\delta\sigma(L) = +\frac{2e^2}{\pi h} \ln \left(\frac{L}{l_e} \right) \tag{2.31}$$

which is the scalling law for WAL.

Fig. 2.6 The contribution from intra and intervalley scattering (Figure is taken from [23])

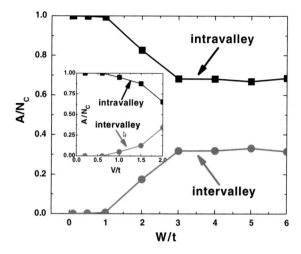

As mentioned above, the Dirac fermions in graphene are expected to exhibit the WAL behavior but another effect should also be involved to consider the whole picture, that is trigonal warping which is related to the momentum contribution from higher order into Eq. (2.15). The trigonal warping is predicted to suppress antilocalization and together with intervalley scattering, it restores the weak localization (WL) [13]. The crossovers from WAL to WL and the effect of disorders on intra- and intervalley scattering were studied in many Refs. [13, 14, 20, 23] in which the long range disorder is simulated by changing onsite energies $V_i = \sum_{j=1}^{N} \epsilon_j \exp[-(\mathbf{r}_i - \mathbf{R}_j)^2/(2\xi^2)]$ where ϵ_j are chosen at random within $\left[-\frac{W}{2}, -\frac{W}{2}\right]$. These calculations show that the strength of local potential profile control the contribution of intra- and intervalley scatterings on the conductivity. Following the theoretical study in Ref. [23], the intravalley scattering dominates at small value of W ($W < \gamma_0$) and valley mixing strength was continuously enhanced from $W = \gamma_0$ to $W = 2\gamma_0$. The intervalley scattering contribution is large enough as $W > 2\gamma_0$ (See. Fig. 2.6). As a consequence, graphene exhibits the crossover from WAL to WL as W increase (See Fig. 2.7). Indeed, the positive magnetoconductance for the case $W = 2\gamma_0$ (top panels) agrees with the strong contribution of intervalley scattering, since all graphene symmetries have been broken. However by decreasing the disorder strength from $W = 2\gamma_0$ to $W = 1.5\gamma_0$ (bottom panels), WAL is indeed recovered given the reduction of intervalley processes.

Chemical absorption in graphene is usually related to oxidation or hydrogenation of graphene which are strongly invasive for electronic and transport properties and systematically drive graphene to a strong Anderson insulator [25]. The theoretical and experimental studies show that high coverage sp^3 formations which break local AB symmetry such as in hydrogenated or fluorinated graphene induce energy band gap in the high density limit. Especially, graphane, fully hydrogenated graphene, is predicted to be a stable semiconductor with the energy gap as large as 3.5 eV [26], some recent DFT calculations using the screened hybrid functional of Heyd, Scuseria,

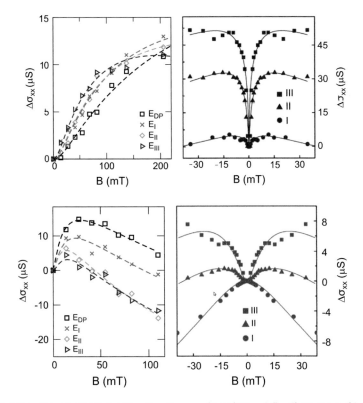

Fig. 2.7 Magnetoconductance for $W = 2\gamma_0$ (*top panels*) and $W = 1.5\gamma_0$ (*bottom panels*), the data is extracted from theoretical (*left panels*) and experimental (*right panels*) study (Figure is taken from [20])

and Ernzerhof (HSE) even gave a larger energy gap up to 4.5 eV for graphane and 5.1 eV for fluorographene (fully fluorinated graphene) (See Fig. 2.8). The case of low coverage of hydrogen is more interesting with the transport properties strongly depending on the absorbing position. Theory predicted that graphene exhibits WL for the compensated case (hydrogen absorbs equally in two sublattices) whereas the quantum interferences and localizations are suppressed if hydrogen defects are restricted to one of the two sublattices [20]. The analogy of transport properties of chemical absorptions and long-range potentials have also been studied. As one can see in Fig. 2.9, some chemical absorptions at bridge position such as epoxide defects which preserve local AB symmetry induce energy-dependent elastic scattering time $(\tau_e(E))$ ressembling the case of long range impurities with small onsite potential depth, whereas some adsorbates at the top position such as hydrogen or fluorine defects which break local sp^2 and AB symmetry give rise to elastic scattering time ressembling the case of strong long range potentials. These are due to the fact that transport time behavior is controled by the contribution of inter- and intravalley scatterings which are mainly determined by the breaking of AB symmetry.

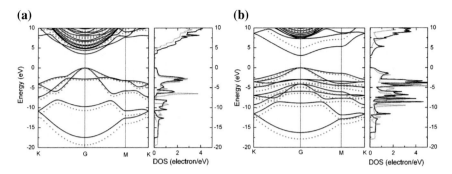

Fig. 2.8 The electronic band structure and projected density of states in the vicinity of the band gap for graphane (**a**) and fluorographene (**b**) (Figure is taken from [24])

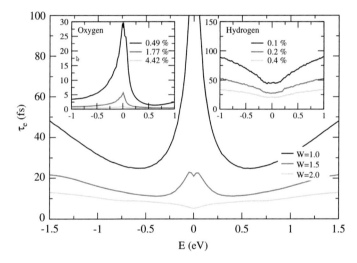

Fig. 2.9 Elastic scattering time (τ_e) versus energy for three different long-range potential strengths W. *Left inset* τ_e for various densities of epoxide defects. *Right inset* τ_e for various densities of hydrogen defects (Figure is taken from [20])

In particular, the formation of sp^3 hybridizations or monovacancies in graphene can give rise to local sublattice imbalances and thus induce local magnetic moment according to Lieb's theorem [27]. The existence of magnetism in graphene as well as magnetism-dependent transport properties have been studied in many Refs. [28–30]. Especially, when half of the hydrogen in graphane sheet is removed, the resulting semihydrogenated graphene (graphone) becomes a ferromagnetic semiconductor with a small indirect gap [31].

Structural point defects usually exist in various geometrical forms in graphene. They can be obtained for instance when irradiating graphene samples. In this kind of graphene, the disorder is created locally in the sample by locally changing the hexagonal structure such as removing a carbon atom from the graphene sheet (monovacancy)

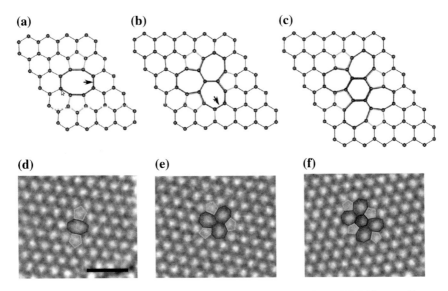

Fig. 2.10 Some structural point defects (*top panels*) and their experimental TEM images (*bottom panels*) (Figure is taken from [35])

or rotating a pair of carbon 90° in graphene plane (Stone-Wales defects). Some studies [32] showed that monovacancies are very mobile and unstable, recombining in di- or multivacancies or local structures with some nonhexagonal rings which are more stable. The transport properties of graphene under the influence of structural point defects such as vacancies, divacancies, Stone-Wales defects, 585 divacancies (See Fig. 2.10), etc. have been now widely studied [33, 34], revealing interesting features such as electron-hole transport asymmetry [33, 34] due to the presence of defect-induced resonances. Under electron irradiation, graphene changes from pristine form to structural defects and finally to a new two-dimensional amorphous carbon lattice [36] which is composed of sp^2-hybridized carbon atoms, arranged as a random tiling of the plane with polygons including four-membered rings. Most theoretical studies [37, 38] found out that there is a huge increase of the density of state at the charge neutrality point in this amorphous graphene and these states are localized, suggesting that the amorphous graphene is an Anderson insulator. However, using a stochastic quenching method, Ref. [39] claimed that "*we predict a transition to metallicity when a sufficient amount of disorder is induced in graphene...*". In Chap. 4, by using Kubo-Greenwood calculations, we show that this conclusion is misleading and similar results have also been obtained recently in Ref. [40].

Although possessing many excellent electrical, optical and mechanical properties, perfect graphene (single-crystal graphene) is only fabricated in small size by exfoliation method. So far, the most promising approach for the mass production of large-area graphene is CVD, which results in a graphene with many line defects (See Fig. 2.11) or Poly-G. This polycrystallinity arises due to the nucleation of growth sites at random positions and orientations during the CVD process. In order to

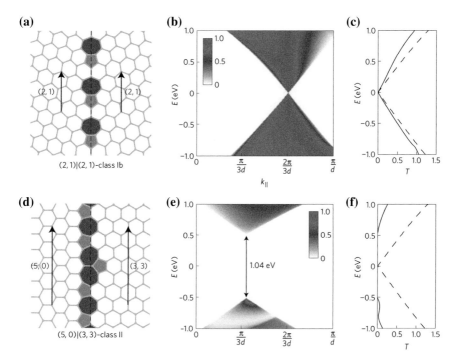

Fig. 2.11 Two classes of electron transport through GBs (Figure is taken from [41])

accommodate the lattice mismatch between misoriented grains, the GBs in Poly-G are made up of a variety of non-hexagonal carbon rings, which can act as a source of scattering during charge transport. Because of its potential for applications, the transport properties of Poly-G are the subject of intense research. Some calculations showed that the effect of GBs on the carrier transport differ depending on the GB geometry (See Fig. 2.11) resulting in a tunable mobility (tunable transport gaps) [41] which allows to control charge currents without the need to introduce bulk band gaps in graphene. In so-called class I GBs (top panels of Fig. 2.11), including all symmetrical GBs, the projected periodicities of the lattice on each side match in a way that allows carriers to cross freely even at the Dirac point. In class II GBs (bottom panels of Fig. 2.11), no such momentum-conserving transmission is possible, except for carriers with much higher energy. Another calculation pointed out that some line defects can play the role as semitransparent "valley filter". It was found that carriers arriving at this line defect with a high angle of incidence are transmitted with a valley polarization near 100 % [42]. Many experimental works have studied the transport properties of Poly-G and showed that the GBs generally degrade the electrical performance of graphene [43, 44] and specifically, the interdomain connectivity plays an important role to control the electrical properties of Poly-G, with the electrical conductance that can be improved by one order of magnitude for GBs with better interdomain connectivity [43]. However, just a few theoretical works have studied the complex structures of GBs and corresponding electronic transport.In Chap. 4, by using molecular dynamics, we simulate the Poly-G with variable grain sizes, and tunable interdomain connectivities, and report on a scaling law for transport properties of Poly-G, which points out that the semiclassical conductivity and mean free path are directly proportional to grain size and both are strongly affected by grain connectivity However, as pointed out in our next calculation, the GB resistivity for non-contaminated Poly-G is very low compared to the experimental results [43, 45, 46]. The explanation for this problem is that the GBs which contain many nonhexagonal structure have greater chemical reactivity [47] and are usually functionalized by many different types of chemical adsorbates. This has been confirmed in several experiments [48, 49]. By using the numerical simulations we report on the role played by chemical adsorbates on GBs in charge transport in Chap. 4.

2.4 Spin Transport in Graphene

Beside many interesting electronic properties, graphene is also considered to be a promising candidate for spintronic applications. The spin relaxation time in intrinsic graphene is expected to be very long and therefore graphene has high potential as a spin-conserver system which can transmit spin-encoded information across a device with high fidelity. The underlying reason for long spin relaxation time is the low hyperfine interactions of the spin with the carbon nuclei (natural carbon only contains 1% ^{13}C) and the weak SOC due to the low atomic number [50]. The theoretical prediction showed that the spin relaxation time in graphene is in the order of microseconds. However, the reported experimental spin relaxation times remain several orders of magnitude lower than the original theoretical predictions.

Because spin relaxation based on the graphene intrinsic SOC could not give a convincing explanation, other extrinsic sources of spin relaxation are believed to come into play. Proposals to explain the unexpectedly short spin relaxation times include spin decoherence due to interactions with the substrate, the extrinsic SOC induced by impurities, adatoms, ripples or corrugations, etc. which will be reviewed below. The puzzling controversy of the spin relaxation mechanism will be mentioned in the next section.

2.4.1 Spin-Orbit Coupling in Graphene

In order to derive the SOC term in the Hamiltonian, it is necessary to start from the relativistic Hamiltonian, the Dirac equation: $H|\psi\rangle = E|\psi\rangle$ with

$$H = \begin{pmatrix} 0 & c\mathbf{p}.\sigma \\ c\mathbf{p}.\sigma & 0 \end{pmatrix} + \begin{pmatrix} mc^2 & 0 \\ 0 & -mc^2 \end{pmatrix} + V \qquad (2.32)$$

and where the wave function is a two-components spinor: $|\psi\rangle = (\psi_A, \psi_B)^T$. From the Dirac equation we obtain two equations for spinor components:

$$\psi_B = \frac{c\mathbf{p}.\sigma}{E - V + mc^2} \psi_A \tag{2.33}$$

$$\mathbf{p}.\sigma \frac{c^2}{E - V + mc^2} \mathbf{p}.\sigma \psi_A = (E - V - mc^2)\psi_A \tag{2.34}$$

In the nonrelativistic limit, the lower component ψ_B is very small compared to the upper component ψ_A. Indeed, with the relativistic energy $E = mc^2 + \epsilon$ and $V \ll mc^2$, Eq. (2.33) drive us to

$$\psi_B = \frac{\mathbf{p}.\sigma}{2mc} \psi_A \ll \psi_A \tag{2.35}$$

and Eq. (2.34) leads us to the Schrodinger equation.[1]

$$\left(\frac{p^2}{2m} + V\right) \psi_A = \epsilon \psi_A \tag{2.36}$$

In other words, in the first order of (v/c), ψ_A is equivalent to the Schrodinger wave function ψ. In order to obtain the analogy of ψ_A and ψ at higher order of (v/c), we use the normalization characteristic of the wave function

$$\int \left(\psi_A^+ \psi_A + \psi_B^+ \psi_B\right) = 1 \tag{2.37}$$

To first order, using Eq. (2.35), this gives

$$\int \psi_A^+ \left(1 + \frac{p^2}{4m^2c^2}\right) \psi_A = 1 \tag{2.38}$$

Apparently, to have a normalized wave function, we should use $\psi = \left(1 + \frac{p^2}{8m^2c^2}\right) \psi_A$. Substituting this into the Dirac equation, and using the expansion $\frac{c^2}{E-V+mc^2} \simeq \frac{1}{2m}\left(1 - \frac{\epsilon-V}{2mc^2} + \cdots\right)$, we obtain, after some rearrangement, the Pauli equation

$$\left(\frac{p^2}{2m} + V - \frac{p^4}{8m^3c^2} - \frac{\hbar}{4m^2c^2}\sigma.\mathbf{p} \times \nabla V + \frac{\hbar^2}{8m^2c^2}\nabla^2 V\right) \psi = \epsilon\psi \tag{2.39}$$

the first and the second terms are the usual terms in the Hamiltonian, the third term is simply a relativistic correction to the kinetic energy. The fourth term is the SOC term and the final term give the energy shift due to the potential.

---

[1] Using $(\sigma.A)(\sigma.B) = \mathbf{A}.\mathbf{B} + i\sigma.(\mathbf{A} \times \mathbf{B})$.

Hereafter, I will derive the SOC term in the more intuitive way which gives the physical meaning of SOC interation. Suppose an electron is moving with velocity $\mathbf{v}$ in an electric field $-e\mathbf{E} = -\nabla V$. This electric field might be induced by the potential V of the adatoms or the substrate. In relativistic theory, this moving electron feels a magnetic field $\mathbf{B} = -\frac{\mathbf{v} \times \mathbf{E}}{c}$ in its rest frame. The interation between this magnetic field and the electron spin leads to the potential energy term:

$$V_{\mu_s} = -\boldsymbol{\mu}_s \mathbf{B} = -\frac{g_s \mu_B}{2ec}\boldsymbol{\sigma}.\mathbf{v} \times \nabla V = -\frac{g_s \hbar}{4m^2c^2}\boldsymbol{\sigma}.\mathbf{p} \times \nabla V = -\frac{\hbar}{2m^2c^2}\boldsymbol{\sigma}.\mathbf{p} \times \nabla V \tag{2.40}$$

This results is twice the SOC term in Pauli equations. Actually, this was the major puzzle, until it was pointed out by Thomas [51] that this argument overlooks a second relativistic effect that is less widely known, but is of the same order of magnitude: electric field $\mathbf{E}$ causes an additional acceleration of the electron perpendicular to its instantaneous velocity $\mathbf{v}$, leading to a curved electron trajectory. In essence, the electron moves in a rotating frame of reference, implying an additional precession of the electron, called the Thomas precession. As a result, the electron "sees" the magnetic field at only one-half the above value

$$\mathbf{B} = -\frac{\mathbf{v} \times \mathbf{E}}{2c} \tag{2.41}$$

which leads to the full SOC term

$$V_{SOC} = -\frac{\hbar}{4m^2c^2}\boldsymbol{\sigma}.\mathbf{p} \times \nabla V \tag{2.42}$$

Now let's rewrite the SOC term in form of the SOC force $\mathbf{F}$

$$H_{SOC} = \alpha\,(\mathbf{F} \times \mathbf{p})\,.\mathbf{s} = -\alpha\,(\mathbf{s} \times \mathbf{p})\,.\mathbf{F} \tag{2.43}$$

where α is an undetermined parameter. Here we use $\mathbf{s}$ instead of $\boldsymbol{\sigma}$ to represent the spin degree of freedom to avoid any misunderstanding with pseudospin in graphene.

If we consider intrinsic graphene, the inversion symmetry dictates the electric field (force) in plane and this SOC is called intrinsic SOC. Because of structure's mirror symmetry with respect to any nearest-neighbor bond (See Fig. 2.12a), the nearest-neighbor intrinsic SOC is zero, while the next nearest-neighbor intrinsic SOC is nonzero. According to symmetry,

$$H_I = i\gamma_2 \left(\mathbf{F}_{//} \times d_{ij}\right).\mathbf{s} = \frac{2i}{\sqrt{3}}V_I\mathbf{s}.(\hat{\mathbf{d}}_{kj} \times \hat{\mathbf{d}}_{ik}) \tag{2.44}$$

where γ_2 and V_I are undetermined parameters, and $\hat{\mathbf{d}}_{ij}$ is the unit vector from atom j two its next-nearest neighbors i, and k is the common nearest neighbor of i and j

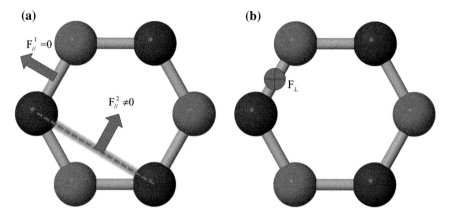

Fig. 2.12 SOC in graphene: **a** Intrinsic SOC forces. **b** Rashba SOC force

In the presence of the out of plane electric field (See Fig. 2.12b) which can originate from a gate voltage or charged impurities in the substrate, adatoms, etc., the band structure of graphene changes. This external electric field breaks spatial inversion symmetry and causes a nearest-neighbor extrinsic SOC. This SOC is Rashba SOC and has the form

$$H_R = i\gamma_1 \left(\mathbf{s} \times \hat{\mathbf{d}}_{ij} \right) . F_\perp \mathbf{e}_z = i V_R \hat{\mathbf{z}}.(\mathbf{s} \times \hat{\mathbf{d}}_{ij}) \tag{2.45}$$

where j is the nearest-neighbor of i and γ_1 and V_R are undetermined parameters.

Finally, we get the TB Hamiltonian:

$$\mathcal{H} = -\gamma_0 \sum_{\langle ij \rangle} c_i^+ c_j + \frac{2i}{\sqrt{3}} V_I \sum_{\langle\langle ij \rangle\rangle} c_i^+ \mathbf{s}.(\hat{\mathbf{d}}_{kj} \times \hat{\mathbf{d}}_{ik}) c_j + i V_R \sum_{\langle ij \rangle} c_i^+ \hat{\mathbf{z}}.(\mathbf{s} \times \hat{\mathbf{d}}_{ij}) c_j \tag{2.46}$$

By performing Fourier transformations, we obtain the low energy effective Hamiltoniam around the Dirac point in the basis $\{|A\rangle, |B\rangle\} \otimes \{|\uparrow\rangle, |\downarrow\rangle\}$

$$h(\mathbf{k}) = h_0(\mathbf{k}) + h_R(\mathbf{k}) + h_I(\mathbf{k}) \tag{2.47}$$

where

$$h_0(\mathbf{k}) = \hbar v_F (\eta \sigma_x k_x + \sigma_y k_y) \otimes 1_s$$
$$h_R(\mathbf{k}) = \lambda_R \left(\eta \left[\sigma_x \otimes s_y \right] - \left[\sigma_y \otimes s_x \right] \right)$$
$$h_I(\mathbf{k}) = \lambda_I \eta \left[\sigma_z \otimes s_z \right] \tag{2.48}$$

with Fermi velocity $v_F = \frac{3}{2}\gamma_0$, Rashba SOC $\lambda_R = \frac{3}{2}V_R$ and intrinsic SOC $\lambda_I = 3\sqrt{3}V_I$ [52].

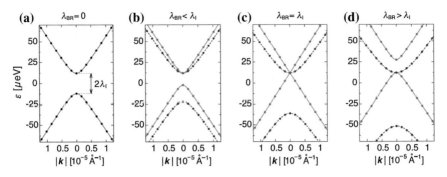

Fig. 2.13 Electronic bandstructure of graphene with SOC (Figure is taken from [53])

The remarkable thing about SOC in graphene is that the SOC terms are momentum-independent. The spin directly couples with pseudospin instead of momentum as in conventional metals or semiconductors, the usual SOC term $(\mathbf{k} \times \mathbf{s})$ is small and can be disregarded.

Diagonalizing the Hamiltonian in Eq. (2.47) gives the electronic bands close to the Dirac point [53, 54]:

$$\epsilon_{\mu\nu}(\mathbf{k}) = \mu\lambda_R + \nu\sqrt{(\hbar v_F k)^2 + (\lambda_R - \lambda_I)^2} \qquad (2.49)$$

where μ and $\nu = \pm 1$ are band indexes.

If we consider intrinsic graphene, the Rashba SOC is vanishingly small, the intrinsic SOC opens a gap $\Delta = 2\lambda_I$ (See Fig. 2.13a). When Rashba SOC is turned on by inversion symmetry breaking (effect from the substrate, the electric field, the corrugations, etc.), the competition of Rashba and intrinsic SOC leads to gap closing. The gap remains finite $\Delta = 2(\lambda_I - \lambda_R)$ for $0 < \lambda_R < \lambda_I$ (Fig. 2.13b). For $\lambda_R > \lambda_I$ the gap closes and the electronic structure is that of a zero gap semiconductor with quadratically dispersing bands (Fig. 2.13d).

The eigenfunctions corresponding to the eigenvalues in Eq. (2.49) are

$$\psi_{\mu\nu}(\mathbf{k}) = \left(\chi_- |\eta e^{-i\eta\varphi} \left[\frac{\epsilon_{\mu\nu} - \lambda_I}{\nu\hbar v_F k} \right]^\eta, 1 \rangle + \mu\chi_+ |{-i}\eta e^{-i(1+\eta)\varphi}, ie^{-i\varphi} \left[\frac{\lambda_I - \epsilon_{\mu\nu}}{\nu\hbar v_F k} \right]^\eta \rangle \right) / C_{\mu\nu}$$

with $\tan\varphi = k_y/k_x$ and the normalization constant [53] $C_{\mu\nu} = \sqrt{2}\left(1 + \left[\frac{\lambda_I - \epsilon_{\mu\nu}}{\hbar v_F k} \right]^{2\eta}\right)^2$. The expectation value of the spin [53, 54],

$$\mathbf{s}_{\mu\nu}(\mathbf{k}) = \frac{\hbar v_F (\mathbf{k} \times \hat{\mathbf{z}})}{\sqrt{(\hbar v_F k)^2 + (\lambda_I - \mu\lambda_R)^2}} = \frac{\hbar v_F k}{\sqrt{(\hbar v_F k)^2 + (\lambda_I - \mu\lambda_R)^2}} \mathbf{n}(\mathbf{k}) \qquad (2.50)$$

where $\mathbf{n(k)} = (sin\varphi, -cos\varphi, 0)$ is the unit vector along the spin direction, called spin vector.

The remarkable characteristic of spin in spin-orbit coupled graphene in Eq. (2.50) is that it is polarized in-plane and perpendicular to electron momentum $\mathbf{k}$. The magnitude of spin polarization $\mathbf{s}$ vanishes when $k \rightarrow 0$. The Chap. 5 will show that these behaviors are due to the fact that spin and pseudospin are strongly coupled close to the Dirac point where the coupling between pseudospin and momentum is zero because of the destructive interference between the three nearest-neighbor hopping paths. And this leads to the spin-pseudospin entanglement, the component of new spin relaxation mechanism that plays a major role in spin relaxation at the Dirac point in ultra clean graphene.

In the case of high energy $\hbar v_F k \gg \lambda_R + \lambda_I$, the pseudospin is mainly controled by momentum via $h_0(\mathbf{k})$ and aligns in the same direction with momentum (in plane). Spin is dictated by pseudospin via $h_R(\mathbf{k})$, as a consequence, spin polarization for a certain momentum in Eq. (2.50) saturates to 1. By successive unitary rotation of $h(\mathbf{k})$ first into the eigenbasis of $h_0(\mathbf{k})$ and then into the spin basis with respect to the direction $\mathbf{n(k)}$ an effective Bychkov-Rashba-type 2×2 Hamiltonian can be obtained for both holes and electrons [9],

$$\tilde{h}(\mathbf{k}) = \nu(\hbar v_F k - \lambda_I) - \nu\lambda_R \mathbf{n(k)}.\mathbf{s} \qquad (2.51)$$

The analogy of the second term in above equation and the original Bychkov-Rashba Hamiltonian in semiconductor heterostructures $H_{\mathbf{k}} = \hbar\mathbf{\Omega(k)}.\mathbf{s}/2$ shows that SOC coupling in graphene effectively acts on the electrons spin as an in-plane magnetic field of constant amplitude but perpendicular to $\mathbf{k}$. In this effective field the spin precesses with a frequency and a period of [9]

$$\Omega = \frac{2\lambda_R}{\hbar}, \qquad T_\Omega = \frac{\pi\hbar}{\lambda_R} \qquad (2.52)$$

These results will be obtained again in Chap. 5 with the numerical calculations of the real-space order N method implemented for spin. Furthermore, we will point out that this result is only valid at high energy. At low energy the spin-pseudospin entanglement comes into play and creates a more complicated picture.

The magnitude of SOC interactions is also a matter of large concern. It is a crucial factor to determine not only quantitatively spin relaxation but also the mechanism at play. The numerical estimates for intrinsic SOC λ_I in graphene remains rather controversial. At the beginning, Kane and Mele [56] estimated the value of 100 μeV. This optimistic estimate was drastically reduced by Min et al. [57] to the value of 0.5 μeV by using microscopic TB model and second-order perturbation theory. This value was later confirmed by Huertas-Hernando et al. [50] with TB model and Yao et al. [58] with first-principles calculations. A density functional calculation of Boettger and Trickey [59], using a Gaussian-type orbital fitting function methodology, gave 2 μeV. Three Refs. [50, 57, 58] gave the same value of λ_I, but these calculations only involved the SOC induced by the coupling of the p_z orbitals (forming

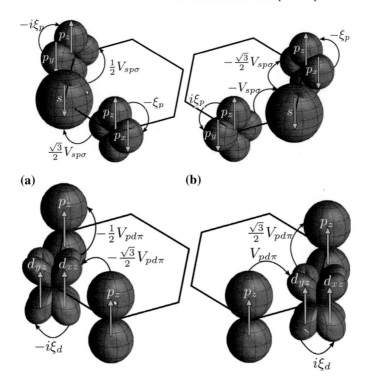

Fig. 2.14 Two possible hopping paths through s and p orbitals (*top panels*) and through d orbital (*bottom panels*) lead to the first and the second terms, respectively in Eq. (2.53) (Figure is taken from [55])

the π bands) to the s orbitals (forming the σ band). However, as pointed out in Ref. [55], the coupling of the p_z orbitals to the d orbitals (See Fig. 2.14) dominates the SOC at $K(K')$. Due to a finite overlap between the neighboring p_z and d_{xz}, d_{yz} orbitals, the intrinsic splitting λ_I is linearly proportional to the spin-orbit splitting of the d states, ξ_d (orbitals higher than d have a smaller overlap and contribute less). In contrast, due to the absence of the direct overlap between the p_z and σ-band orbitals, the usually considered spin-orbit splitting [50, 57, 58] induced by the $\sigma - \pi$ mixing depends only quadratically on the spin-orbit splitting of the p_z orbital, ξ_p, giving a negligible contribution.

$$\lambda_I \simeq \frac{2(\varepsilon_p - \varepsilon_s)}{9V_{sp\sigma}^2}\xi_p^2 + \frac{9V_{pd\pi}^2}{2(\varepsilon_d - \varepsilon_p)^2}\xi_d \qquad (2.53)$$

where $\varepsilon_{s,p,d}$ are the energies of s, p, d orbitals, respectively and $V_{sp\sigma}$ and $V_{pd\pi}$ are hopping parameters of the p orbital to the s and d orbital, respectively (Fig. 2.14).

This TB calculation gave the value of intrinsic SOC $\lambda_I = 12\,\mu eV$ [55] and was confirmed by the first principle calculation [53]. These calculations also showed that

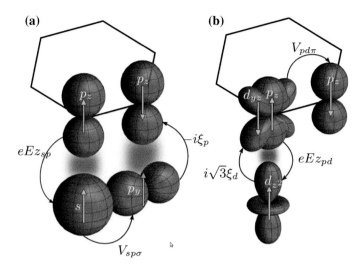

Fig. 2.15 A representative hopping path is responsible for Rashba SOC in Eq. (2.54) (Figure is taken from [55])

the Rashba term (zero in absence of electric field) is tunable with an external electric field E which is perpendicular to graphene plane (Fig. 2.15)

$$\lambda_R \simeq \frac{2eEz_{sp}}{3V_{sp\sigma}}\xi_p + \sqrt{3}\frac{eEz_{sp}}{(\varepsilon_d - \varepsilon_p)}\frac{3V_{pd\pi}}{(\varepsilon_d - \varepsilon_p)}\xi_d \qquad (2.54)$$

where z_{sp} and z_{pd} are the expectation values $\langle s|\hat{\mathbf{z}}|p_z\rangle$ and $\langle p_z|\hat{z}|d_{z^2}\rangle$, respectively, of the operator $\hat{z}$.

All these calculations predicted that the Rashba SOC is directly proportional to the electric field E, but the estimated values vary by about an order of magnitude from $5\,\mu$eV in Ref. [53] to $47\,\mu$eV in Ref. [50] and to $67\,\mu$eV in Ref. [57], for a typical electric field of $E = 1\,$V/nm. Furthermore, Ast and Gierz [60] used the TB model and directly considered the nearest-neighbor contribution from the electric field and obtained $\lambda_R = 37.4\,\mu$eV.

In general, the intrinsic SOC of graphene is very weak, in the order of μeV and is unmeasurable. This makes some phenomena such as QSH effect unobservable in graphene, the material in which it was originally predicted [56]. A way to observe QSH effect in graphene is endowing it with heavy adatoms which increase SOC in graphene. This problem will be mentioned in Chap. 5.

2.4.2 Spin Transport in Graphene

The graphene SOC in the order of μeV as mentioned above should lead to spin relaxation times in the microsecond scale [9]. However, the experimental results is in the order of nanoseconds, several orders of magnitude lower than the original theoretical prediction. In order to clarify the limitations and mechanisms for spin relaxation in graphene a lot of effort has been done by both experimentalists and theoreticians, but up to now this topic is still under debate. The first measurement of electron spin relaxation was performed by Tombros et al. [2] using the non-local spin valve measurement and Hanle spin precession method to study spin relaxation in mechanical exfoliated single-layer graphene (SLG) on SiO_2 substrate with mobility of the devices about 2,000 $cm^2V^{-1}s^{-1}$. They extracted the spin relaxation time of few hundreds of ps and spin relaxation length of few μm at room temperature, similar to what one might expect for conventional metals or semiconductors. This value has been confirmed by several measurements [7, 61]. The spin transport was found to be relatively insensitive to the temperature and weakly dependent on the direction of spin injection and charge density. Due to the fast spin relaxation was attributed to the extrinsic SOC in the substrate and the way to grow graphene, spin measurements in many other kinds of graphene and substrates have been reported. The measurement of spin relaxation on epitaxially grown graphene on $SiC(0001)$ [62] is the first report of spin transport in graphene on a different substrate than SiO_2. The value of spin relaxation τ_s was obtained in the order of few nanoseconds, one order of magnitude larger than in exfoliated graphene on SiO_2. However the spin diffusion coefficient $D_s \approx 4\,cm^2/s$ is about 80 times smaller, yielding to 70 % lower value for spin relaxation length λ_s. The longer τ_s but much smaller D_s was later explained by the influence of localized states arising from the buffer layer at the interface between the graphene and the SiC surface that couple to the spin transport channel [63]. The measurement also reported that τ_s is weakly influenced by the temperature with reductions of D_s by more than 40 % and τ_s by about 20 % at room temperature. With the expectation that removing the underneath substrate helps to reduce the extrinsic SOC and leads to long spin relaxation time, the spin measurement on suspended graphene was performed [64]. Although a high mobility $\mu \approx 10^5\,cm^2V^{-1}s^{-1}$, an increase up to an order of magnitude in spin diffusion coefficient ($D_s = 0.1m^2/s$) compared to SiO_2 supported graphene and long mean free path in the order of a μm were observed, indicating that much less scattering happens, the spin relaxation time remains a few hundreds of ps and spin relaxation length few μm. Other group used CVD method to grow graphene on copper (Cu) substrate and studied the effect of corrugation on spin relaxation time [65]. They observed the same spin relaxation time as in exfoliated graphene and showed that ripples in graphene flakes have minor effects on spin transport parameters.

The nature of spin relaxation is actually a fundamental debated issue. The DP [1, 66, 67] and the EY (EY) [68, 69] are two mechanisms usually discussed in the context of graphene. The EY mechanism is a suitable mechanism for spin relaxation in metals. In the EY mechanism, electron spin changes its direction during the scattering event thanks to the SOC which produces admixtures of spin and electron momentum in the wave functions. Due to these admixtures, scattering changes electron momentum and induces spin-flip probability at the same time and leads to a typical scaling behavior of spin relaxation time with momentum relaxation time $\tau_s^{EY} \sim \tau_p$. On the other hand, DP mechanism is an efficient mechanism for materials with broken inversion symmetry. In these kinds of materials, SOC induces an effective momentum-dependent magnetic field about which electron spin precesses between scattering events. The longer time electron travels, the larger angle electron spin precesses and as a consequence, the more spin dephasing between electrons in the ensemble is accumulated. Therefore, spin relaxation time is inversely proportional to elastic scattering time $\tau_s^{DP} \sim \tau_p^{-1}$. W. Han and R.K. Kawakami performed systematic studies of spin relaxation in SLG and bilayer graphene (BLG) spin valves with tunneling contact [61]. They found that in SLG, the spin relaxation time varies linearly with momentum scattering time τ_p, indicating the dominance of EY spin relaxation whereas in BLG, τ_s and τ_p exhibit an inverse dependence, which indicates the dominance of DP mechanism. However, Pi et al. reported a surprising result that τ_s increases with decreasing τ_p in the surface chemical doping experiment with Au atoms on graphene [5], indicating that the DP mechanism is important there. This experiment led to the conclusion that charged impurity scattering is not the dominant mechanism for spin relaxation, despite its importance for momentum scattering. Even more puzzling, Zomer et al. [70] performed spin transport measurements on graphene deposited on boron nitride with mobilities up to $4.10^4 \, \mathrm{cm^2 V^{-1} s^{-1}}$ and showed that neither EY nor DP mechanisms alone allow for a fully consistent description of spin relaxation. Furthermore, electron spin is expected to relaxe faster in BLG than in SLG because the SOC in BLG is one order of magnitude larger than the one in SLG due to the mixing of π and σ bands by interlayer hopping [71], but the experimental results showed an opposite behavior [61, 72]. The spin relaxation time in BLG has been reported in the order of few *nanoseconds* and show the dominance of DP spin scattering [61, 72]. Now we look at both mechanisms in more detail.

DP mechanism:

As one can see from Eq. (2.51), electron precesses about the effective magnetic field in plane $\mathbf{B}_{\|}(\mathbf{k}) \sim \mathbf{\Omega}(\mathbf{k})$ between scattering events. Random scattering induces motional narrowing of this spin precession causing spin relaxation (See Fig. 2.16). The spin relaxation rates for the α-th spin component following the DP mechanism are [9]

$$\frac{1}{\tau_{s,\alpha}^{DP}} = \tau^* \left(\left\langle \mathbf{\Omega}^2(\mathbf{k}) \right\rangle - \left\langle \mathbf{\Omega}_\alpha^2(\mathbf{k}) \right\rangle \right) \tag{2.55}$$

where τ^* is the correlation time of the random spin-orbit field. In graphene this value coincides with momentum relaxation time $\tau^* = \tau_p$ [9, 73] and the symbol $\langle \cdots \rangle$ expresses an average over the Fermi surface. Because of $\langle \mathbf{\Omega}^2(\mathbf{k}) \rangle = (2\lambda_R/\hbar)^2$, $\langle \mathbf{\Omega}_z^2(\mathbf{k}) \rangle = 0$ and $\langle \mathbf{\Omega}_{x,y}^2(\mathbf{k}) \rangle = \frac{1}{2}(2\lambda_R/\hbar)^2$, the DP relation for spin relaxation in graphene is [1, 9]

$$\tau_{s,z}^{DP} = \frac{\hbar^2}{4\lambda_R^2 \tau_p}, \quad \text{and} \quad \tau_{s,\{x,y\}}^{DP} = 2\tau_{s,z}^{DP} = \frac{\hbar^2}{2\lambda_R^2 \tau_p} \tag{2.56}$$

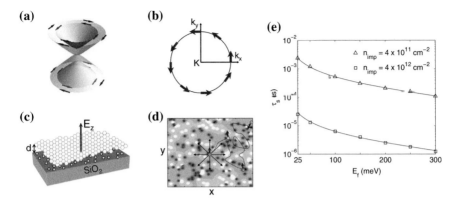

Fig. 2.16 DP spin relaxation in graphene: **a** Dirac coin when SOC is included. **b** $B_\parallel(k)$ along the Fermi circle. **c** Charged impurities in substrate induce electric field in graphene. **d** Illustration of the spin relaxation in a spatially random potential due to the charged carriers. **e** Calculated spin relaxation time τ_s as a function of the Fermi energy E_f (Figure is taken from [9])

Because the spin relaxation time is inversely proportional to the momentum relaxation time, the DP spin relaxation length is independent of mean free path [1].

$$\lambda_s = \sqrt{D\tau_s} = \sqrt{\frac{1}{2}v_F^2 \tau_p \tau_s} = \frac{\hbar v_F}{2\sqrt{2}\lambda_R} \qquad (2.57)$$

The analytical estimates and Monte Carlo simulations [9] with DP mechanism show that the corresponding spin relaxation times are between micro- to milliseconds (See Fig. 2.16) several orders of magnitude larger then the experimental results.

EY mechanism:

As mentioned above, intrinsic SOC obtained by TB model and density functional calculation is in the order of tens μeV [50, 55, 57, 58], much smaller and can be neglected in comparison to the Rashba SOC. In the case of slowly varied Rashba SOC induced by electric field or ripples, the Hamiltonian can be written in form

$$\mathcal{H} = -i\hbar v_F \boldsymbol{\sigma}.\nabla + \lambda_R(\boldsymbol{\sigma} \times s) \qquad (2.58)$$

Because of the Rashba SOC, Bloch states with well-defined spin polarization are no longer eigenstates of the Hamiltonian. The Bloch eigenstates of above Hamiltonian are [69]

$$\Psi_{k,\pm} = \left[\left(\frac{1}{\frac{\epsilon_{k\pm}}{\hbar v_F k}e^{i\theta_k}} \right) \otimes | \uparrow \rangle \pm i \left(\frac{\frac{\epsilon_{k\pm}}{\hbar v_F k}e^{i\theta_k}}{e^{2i\theta_k}} \right) \otimes | \downarrow \rangle \right] e^{ikr}. \qquad (2.59)$$

where $\theta_k = arctan(k_y/k_x)$ and the energy $\epsilon_{k\pm} = \pm\lambda_R + \sqrt{(\hbar v_F k)^2 + \lambda_R^2}$ is obtained from Eq. (2.49) with $\lambda_I = 0$. When $\lambda_R = 0$, eigenstates in Eq. (2.59)

Fig. 2.17 Sketch of scattering by a potential $U(\mathbf{r})$ in the chiral channels (Figure is taken from [69])

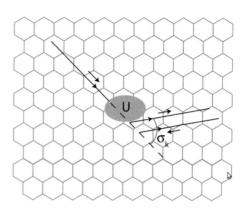

have their spin pointing along (helicity $+$) or opposite to (helicity $-$) the direction of motion. This is not true when $\lambda_R \neq 0$ but in the case of $\lambda_R/\epsilon_F \ll 1$, using perturbation theory we can identify each of these eigenstates with chiral states $\pm$ [69]. Let's consider the Born approximation of the scattering problem of electron in the graphene under the local scattering potential $U(\mathbf{r})$ which is diagonal in the sublattice and spin degrees of freedom. The scattering amplitudes $f_\pm^0(\theta)$ for chiral channels $\pm$ of an incoming electron with positive chirality in the case of $\lambda_R = 0$ are (For detail derivation, see Ref. [69])

$$f_+^0(\theta) = -(\hbar v_F)^{-1}\sqrt{\tfrac{k}{8\pi}} U_{\mathbf{q}} e^{-i\theta}(1 + cos\theta)$$

$$f_-^0(\theta) = -(\hbar v_F)^{-1}\sqrt{\tfrac{k}{8\pi}} U_{\mathbf{q}} i e^{-i\theta} sin\theta \qquad (2.60)$$

where $U_{\mathbf{q}}$ is the Fourier transformation of the scattering potential evaluated at the transferred momentum $\mathbf{q} = \mathbf{k}' - \mathbf{k}$ and angle θ (see Fig. 2.17) between the outgoing momentum $\mathbf{k}'$ and incoming momentum $\mathbf{k}$.

When Rashba SOC is turned on these amplitudes become

$$f_+^{\lambda_R}(\theta) = -(\hbar v_F)^{-2}\sqrt{\tfrac{1}{8\pi k_+}}\,(\epsilon + (\epsilon - 2\lambda_R)cos\theta)\, U_{\mathbf{q}_+} e^{-i\theta}$$

$$f_-^{\lambda_R}(\theta) = -(\hbar v_F)^{-2}\sqrt{\tfrac{1}{8\pi k_-}}\,(\epsilon + 2\lambda_R)\, U_{\mathbf{q}_-} i e^{-i\theta} sin\theta \qquad (2.61)$$

where $k_\pm = (\hbar v_F)^{-1}\sqrt{\epsilon^2 \mp 2\epsilon\lambda_R}$ and $\mathbf{q}_\pm = \mathbf{k}'_\pm - \mathbf{k}$.

Let us define the probability for a spin-flip process from the changes in the scattering in both chiral channels due to the presence of the SOC.

$$S(\theta) = \frac{\sum_{\pm 1} |f_\pm^0(\theta)| |f_\pm^{\lambda_R}(\theta) - f_\pm^0(\theta)|}{\sum_{\pm 1} |f_\pm^0(\theta)|^2} \qquad (2.62)$$

This is the amount of spin relaxed in the direction defined by θ. The total amount of spin relaxation during a scattering event can be defined as the average of this quantity over the Fermi surface:

$$S = \langle S(\theta) \rangle = \frac{1}{2\pi} \int d\theta S(\theta, \epsilon = \epsilon_F) \tag{2.63}$$

It is easy to see that $f_{\pm}^{\lambda_R}(\theta) - f_{\pm}^{0}(\theta) \sim \lambda_R/\epsilon_F$ from expanding of Eq. (2.61) in powers of λ_R/ϵ_F. This implies that $S(\theta) \sim \lambda_R/\epsilon_F$ which is independent with the scattering potentials $U(\mathbf{r})$. This result was obtained in Ref. [1] for the case of weak scatterers, and later in Ref. [69] for the cases of scattering by boundary, strong scatterers and clusters of impurities which can not be treated in perturbation theory. Assuming this behavior, the EY relation for graphene can be easily found. Indeed, the change of spin orientation at each collision is $S \sim \lambda_R/\epsilon_F$. The total change of spin orientation after N_{col} collisions is of the order of $\sqrt{N_{col}}\epsilon_F/\lambda_R$. Dephasing occurs when $\sqrt{N_{col}}\epsilon_F/\lambda_R \sim 1$ and of course, after a time $\tau_s^{EY} = N_{col}\tau_p$. Hence we obtain the EY relation

$$\tau_s^{EY} \approx \frac{\epsilon_F^2}{\lambda_R^2}\tau_p \tag{2.64}$$

This is the EY relation for graphene. It is worth to mention that the spin relaxation time τ_s here not only is proportional to momentum relaxation time τ_p but also depends on the carrier density through Fermi energy ϵ_F. The spin relaxation length in EY mechanism is proportional to mean free path ℓ_e

$$\lambda_s = \sqrt{D\tau_s} = \sqrt{\frac{1}{2}v_F^2\tau_p\tau_s} \sim \ell_e\frac{\epsilon_F}{\sqrt{2}\lambda_R} \tag{2.65}$$

Despite the fact that some experiments have reported that $\tau_s \sim \tau_p$, indicating the dominance of EY mechanism in spin relaxation in graphene, the discrepancy between theoretical calculations and experimental data is still large. Furthermore, the derivations of both EY and DP for graphene are based on the strong coupling of momentum and pseudospin which is unsuitable close to the Dirac point. In Chap. 5, we propose a new mechanism which is the heart of this PhD thesis to explains the fast spin relaxation in graphene by the entanglement of spin and pseudospin degrees of freedom.

References

1. D. Huertas-Hernando, F. Guinea, A. Brataas, Phys. Rev. Lett. **103**, 146801 (2009)
2. N. Tombros et al., Nature (London) **448**, 571 (2007)
3. N. Tombros et al., Phys. Rev. Lett. **101**, 046601 (2008)
4. C. Jozsa et al., Phys. Rev. Lett. **100**, 236603 (2008)
5. K. Pi et al., Phys. Rev. Lett. **104**, 187201 (2010)

6. T. Maassen et al., Phys. Rev. B **83**, 115410 (2011)
7. W. Han et al., Phys. Rev. Lett. **105**, 167202 (2010)
8. A.H. Castro Neto, F. Guinea, N.M.R. Peres, K.S. Novoselov, A.K. Geim, Rev. Mod. Phys. **81**, 109 (2009)
9. C. Ertler et al., Phys. Rev. B **80**, 041405 (2009)
10. E.F. Luis, F. Torres, S. Roche, J.C. Charlier, *Introduction to Graphene-Based Nanomaterials From Electronic Structure to Quantum Transport* (Cambridge, 2013)
11. S.D. Sarma, S. Adam, E.H. Hwang, E. Rossi, Rev. Mod. Phys. **83**, 407 (2011)
12. M.O. Goerbig, Rev. Mod. Phys. **83**, 1193 (2011)
13. E. McCann, K. Kechedzhi, V.I. Falko, H. Suzuura, T. Ando, B.L. Altshuler, Phys. Rev. Lett. **97**, 146805 (2006)
14. F.V. Tikhonenko, D.W. Horsell, R.V. Gorbachev, A.K. Savchenko, Phys. Rev. Lett. **100**, 056802 (2008)
15. M.I. Katsnelson, K.S. Novoselov, A.K. Geim, Nat. Phys. **2**, 620 (2006)
16. K.S. Novoselov, A.K. Geim, S.V. Morozov, D. Jiang, Y. Zhang, S.V. Dubonos, I.V. Grigorieva, A.A. Firsov, Science **306**, 666 (2004)
17. K.S. Novoselov, A.K. Geim, S.V. Morozov, D. Jiang, M.I. Katsnelson, I.V. Grigorieva, S.V. Dubonos, A.A. Firsov, Nature (London) **438**, 197 (2005)
18. A.K. Geim, K.S. Novoselov, Nat. Mater. **6**, 183 (2007)
19. N.H. Shon, T. Ando, J. Phys. Soc. Jpn. **67**, 2421 (1998)
20. S. Roche, N. Leconte, F. Ortmann, A. Lherbier, David Soriano, Jean-Christophe Charlier, Solid State Commun. **152**, 1404–1410 (2012)
21. A. Lherbier, B. Biel, Y.M. Niquet, S. Roche, Phys. Rev. Lett. **100**, 036803 (2008)
22. P.A. Lee, T.V. Ramakrishnan, Rev. Mod. Phys. **57**, 2 (1985)
23. Y.Y. Zhang, J. Hu, B.A. Bernevig, X.R. Wan, X.C. Xie, W.M. Liu, Phys. Rev. Lett. **102**, 106401 (2009)
24. F. Karlicky, R. Zboil, M. Otyepka, J. Chem. Phys. **137**, 034709 (2012)
25. F. Evers, A.D. Mirlin, Rev. Mod. Phys. **80**, 1355 (2008)
26. J.O. Sofo, A.S. Chaudhari, G.D. Barber, Phys. Rev. B **75**, 153401 (2007)
27. E.H. Lieb, Phys. Rev. Lett. **62**, 1201 (1989)
28. V.O. Yazyev, Phys. Rev. Lett. **101**, 037203 (2008)
29. D. Soriano, N. Leconte, P. Ordejon, J.C. Charlier, Juan J. Palacios, Stephan Roche, Phys. Rev. Lett. **107**, 016602 (2011)
30. N. Leconte, D. Soriano, S. Roche, P. Ordejon, Jean C. Charlier, Juan J. Palacios, ACS Nano **5**, 3987–3992 (2011)
31. J. Zhou et al., Nano Lett. **9**, 3867–3870 (2009)
32. A.W. Robertson et al., Nat. Commun. **3**, 1144 (2012)
33. A. Lherbier, S.M.M. Dubois, X. Declerck, S. Roche, Y.M. Niquet, J.C. Charlier, Phys. Rev. Lett. **106**, 046803 (2011)
34. A. Lherbier, S.M.M. Dubois, X. Declerck, Y.M. Niquet, S. Roche, J.C. Charlier, Phys. Rev. B **86**, 075402 (2012)
35. F. Banhart, J. Kotakoski, A.V. Krasheninnikov, ACS Nano **5**, 26–41 (2011)
36. J. Kotakoski, A.V. Krasheninnikov, U. Kaiser, J.C. Meyer, Phys. Rev. Lett. **106**, 105505 (2011)
37. V. Kapko, D.A. Drabold, M.F. Thorpe, Phys. Stat. Solidi B **247**, 1197–1200 (2010)
38. Y. Li, F. Inam, F. Kumar, M.F. Thorpe, D.A. Drabold, Phys. Stat. Solidi B **248**, 2082–2086 (2011)
39. E. Holmström et al., Phys. Rev. B **84**, 205414 (2011)
40. A. Lherbier, S. Roche, O.A. Restrepo, Y.-M. Niquet, A. Delcorte, J.-C. Charlier, Nano Res. **6**, 326–334 (2013)
41. O.V. Yazyev, S.G. Louie, Nat. Mater. **9**, 806 (2010)
42. D. Gunlycke, C.T. White, Phys. Rev. Lett. **106**, 136806 (2011)
43. A.W. Tsen et al., Science **336**, 1143–1146 (2012)
44. P.Y. Huang, C.S. Ruiz-Vargas, A.M. van der Zande, W.S. Whitney, M.P. Levendorf, J.W. Kevek, S. Garg, J.S. Alden, C.J. Hustedt, Y. Zhu, J. Park, P.L. McEuen, D.A. Muller, Nature **469**, 389 (2011)

45. Q. Yu et al., Nat. Mater. **10**, 443–449 (2011)
46. K.W. Clark et al., ACS Nano **7**, 7956–7966 (2013)
47. D.W. Boukhvalov, M.I. Katsnelson, Nano Lett. **8**, 4373–4379 (2008)
48. P. Nemes-Incze et al., Appl. Phys. Lett. **94**, 023104 (2011)
49. D.L. Duong et al., Nature **490**, 235–239 (2012)
50. D. Huertas-Hernando, F. Guinea, A. Brataas, Phys. Rev. B **74**, 155426 (2006)
51. L.H Thomas, Nature **117**, 514 (1926)
52. Z. Qiao, H. Jiang, X. Li, Y. Yao, Q. Niu, Phys. Rev. B **85**, 115439 (2012)
53. M. Gmitra, S. Konschuh, C. Ertler, C. Ambrosch-Draxl, J. Fabian. Phys. Rev. B **80**, 235431 (2009)
54. I. Emmanuel Rashba, Phys. Rev. B **79**, 161409(R) (2009)
55. S. Konschuh, M. Gmitra, J. Fabian. Phys. Rev. B **82**, 245412 (2010)
56. C.L. Kane, E.J. Mele, Phys. Rev. Lett. **95**, 226801 (2005)
57. H. Min et al., Phys. Rev. B **74**, 165310 (2006)
58. Y. Yao et al., Phys. Rev. B **75**, 041401(R) (2007)
59. J.C. Boettger, S.B. Trickey, Phys. Rev. B **75**, 121402(R) (2007)
60. C.R. Ast, I. Gierz, Phys. Rev. B **86**, 085105 (2012)
61. W. Han, R.K. Kawakami, Phys. Rev. Lett. **107**, 047207 (2011)
62. T. Maassen et al., Nano Lett. **12**, 1498–1502 (2012)
63. T. Maassen, J.J. van den Berg, E.H. Huisman, H. Dijkstra, F. Fromm, T. Seyller, B.J. van Wees, Phys. Rev. Lett. **110**, 067209 (2013)
64. M.H.D. Guimaraes, A. Veligura, P.J. Zomer, T. Maassen, I.J. Vera-Marun, N. Tombros, B.J. van Wees, Nano Lett. **12**, 3512–3517 (2012)
65. A. Avsar et al., Nano Lett. **11**, 2363–2368 (2011)
66. M.I. D'yakonov, V.I. Perel, Zh. Eksp. Teor. Fiz. **60**, 1954 (1971)
67. M.I. D'yakonov, V.I. Perel, Sov. Phys. Solid State **13**, 3023 (1971)
68. R.J. Elliot, Phys. Rev. **96**, 266 (1954)
69. H. Ochoa, A.H. Castro, Neto, F. Guinea, Phys. Rev. Lett. **108**, 206808 (2012)
70. P.J. Zomer, M.H.D. Guimaraes, N. Tombros, B.J. van Wees, Phys. Rev. B **86**, 161416(R) (2012)
71. F. Guinea, New J. Phys. **12**, 083063 (2010)
72. T.Y. Yang et al., Phys. Rev. Lett. **107**, 047206 (2011)
73. J. Fabian, A. Matos-Abiague, C. Ertler, P. Stano, I. Zutic, Acta Phys. Slov. **57**, 565 (2007)

Chapter 3
The Real Space Order $O(N)$ Transport Formalism

Quantums simulations are very important tools to study transport phenomena in the nanoscale. There are two numerical approaches for quantum transport simulations at the present, one is the widely used non-equilibrium Green's function (NEGF) method, the other is the Kubo-Greenwood method. While NEGF is usually used for small systems such as carbon nanotubes (CNT), graphene nanoribons (GNRs), due to the cubic-scaling time consumption, the linear-scaling Kubo-Greenwood quantum transport simulation method is a very effective method to investigate the transport properties of the large-scale disorder systems.

In this chapter, the theoretical background of Kubo-Greenwood formalism, the real-space Kubo formulas for conductivity and the Einstein relations are derived in the first section. At the end of this section the three different regimes of transport is discussed. In the second section, a new formalism basing on the real space order $O(N)$ is firstly developed to study the spin transport in large scale 2D system. This method is applied in Chap. 5 to study the spin transport in disordered graphene.

3.1 Electrical Transport Formalism

3.1.1 Electrical Resistivity and Conductivity

When there is a electric field $\mathbf{E}(\omega)$ inside a material, it will cause electric current to flow. Electrical resistivity $\rho(\omega)$ is a measure of how strongly a material opposes the flow of electric current. A low resistivity indicates a material that readily allows the movement of electric charge. The electrical resistivity is defined as the ratio of the electric field to the density of the current $\mathbf{j}(\omega)$, implying

$$\mathbf{j}(\omega) = \frac{\mathbf{E}(\omega)}{\rho(\omega)} \tag{3.1}$$

© Springer International Publishing Switzerland 2016
D.V. Tuan, *Charge and Spin Transport in Disordered Graphene-Based Materials*, Springer Theses, DOI 10.1007/978-3-319-25571-2_3

Conductivity $\sigma(\omega)$ is the inverse of resistivity

$$\sigma(\omega) = \frac{1}{\rho(\omega)} \tag{3.2}$$

Therefore, we have

$$\mathbf{j}(\omega) = \sigma(\omega)\mathbf{E}(\omega) \tag{3.3}$$

We usually measure the response of system to the electric field along 1 direction (ex. the x direction). In this case, the conductivity σ (more specifically σ_{xx}) in this direction is:

$$j_x(\omega) = \sigma(\omega)E_x(\omega) \tag{3.4}$$

$\sigma(\omega)$ is the conductivity in the general case. For the direct current (DC), the DC conductivity σ_{DC} can be obtain by setting $\omega \to 0$.

3.1.2 Semiclassical Approach

Firstly, let's use the semiclassical approach to have the general picture of the motion of electron in the system under the influence of electric field. In this section some formulas such as Drude conductivity, Einstein relation, and Landauer formula are derived which will give a better vision of the quantum approach which will be given in the next section.

In the presence of an electric field $\mathbf{E}$ in the plane of the two dimensional electron gas (2DEG) system, beside thermal motion, electron moves along the direction of electric force. However, this drift motion only remains for a short time before its direction is randomized by scattering on disorder. An electron acquires a drift velocity $\mathbf{v}_{drift} = -e\mathbf{E}\Delta t/m$ in the time Δt since the last impurity collision. The average of Δt is the scattering time τ_p (or momentum relaxation time), so the average drift velocity $\mathbf{v}_{drift}$ is given by [1]

$$v_{drift} = -\mu_e E, \qquad \mu_e = \frac{e\tau_p}{m} \tag{3.5}$$

where μ_e is electron mobility. If the sheet density is n_s then the current density is

$$\mathbf{j} = -en_s\mathbf{v}_{drift} = \sigma\mathbf{E} \tag{3.6}$$

The result is the familiar Drude conductivity [2] which can be written in several equivalent forms:

$$\sigma = en_s\mu_e = \frac{en_s\tau_p}{m} = g_s g_v \frac{e^2}{h} \frac{k_F l}{2} \tag{3.7}$$

In the last equality we have used the identity $n_s = g_s g_v k_F^2 / 4\pi$ which is true for all 2DEG system including graphene [3] and have defined the mean free path $\ell_e = v_F \tau_p$. The valley degeneracy factors are typically $g_v = 2$ for graphene (K and K') and Si 100 based 2DEG system, whereas $g_v = 1(6)$ for 2DEG system in GaAs (Si 111). The spin degeneracy is always $g_s = 2$, except at high magnetic fields.

It is obvious that the current induced by the applied electric field is carried by all conduction electrons, since each electron acquires the same average drift velocity. Nonetheless, to determine the conductivity it is sufficient to consider the response of electrons near the Fermi level to the electric field. The reason is that the states that are more than a few times the thermal energy $k_B T$ below E_F are all filled so that, in response to a weak electric field, only the distribution of electrons among states at energies close to E_F is changed from the equilibrium Fermi-Dirac distribution

$$f(E - E_F) = \left(1 + e^{\frac{E - E_F}{k_B T}}\right)^{-1} \tag{3.8}$$

In thermodynamic equilibrium at zero temperature, which is characterized by a spatially constant electrochemical potential μ, the sum of the drift current density $-\sigma \mathbf{E}/e$ and the diffusion current density $-D \nabla n_s$ (D is diffusion constant) vanishes

$$-\sigma \mathbf{E}/e - D \nabla n_s = 0 \qquad \text{when} \quad \nabla \mu = 0 \tag{3.9}$$

The electrochemical potential μ is the sum of the electrostatic potential energy $-eV$ and the chemical potential at E_F. Since $dE_F/dn_s = 1/\rho(E_F)$, one has

$$\nabla \mu = -e\nabla V + \nabla E_F = e\mathbf{E} + \frac{\nabla n_s}{\rho(E_F)} \tag{3.10}$$

The combination of Eqs. (3.9) and (3.10) yields the Einstein relation for the conductivity σ

$$\sigma = e^2 \rho(E_F) D \tag{3.11}$$

To verify that Eq. (3.11) is consistent with the earlier expression (3.7) for the Drude conductivity, one can use the result (see below) for the 2D diffusion constant:

$$D = \frac{1}{2} v_F^2 \tau_p = \frac{1}{2} v_F l \tag{3.12}$$

in combination with the density of states: $\rho(E) = g_s g_v E / 2\pi (\hbar v_F)^2$ for graphene and $\rho(E) = g_s g_v m / 2\pi \hbar^2$ for 2DEG systems.

The result in Eq. (3.12) can be explained in the following way [1]. Consider the diffusion current density j_x induced by a small constant density gradient, $n(x) = n_0 + cx$. We write

$$
\begin{aligned}
j_x &= \lim_{\Delta t \to \infty} \langle v_x(t=0)n(x(t=-\Delta t)) \rangle \\
&= \lim_{\Delta t \to \infty} c \langle v_x(0)x(-\Delta t) \rangle \\
&= \lim_{\Delta t \to \infty} -c \int_0^{\Delta t} dt \langle v_x(0)v_x(-t) \rangle
\end{aligned}
$$

where the brackets $\langle ... \rangle$ denote an isotropic angular average over the Fermi surface. The time interval $\Delta t \to \infty$, so the velocity of the electron at time 0 is uncorrelated with its velocity at the earlier time $-\Delta t$. This allows us to neglect at $x(-\Delta t)$ the small deviations from an isotropic velocity distribution induced by the density gradient [which could not have been neglected at $x(0)$]. Since only the time difference matters in the velocity correlation function, one has $\langle v_x(0)v_x(-t) \rangle = \langle v_x(t)v_x(0) \rangle$. We thus obtain for the diffusion constant $D = -j_x/c$ the familiar linear response formula [4]

$$
D = \int_0^\infty dt \langle v_x(t)v_x(0) \rangle \tag{3.13}
$$

Since, in the semiclassical relaxation time approximation, each scattering event is assumed to destroy all correlations in the velocity, and since a fraction e^{-t/τ_p} of the electrons has not been scattered in a time t, one has (in 2D) the result in Eq. (3.12)

$$
D = \int_0^\infty dt \left\langle v_x(0)^2 \right\rangle e^{-t/\tau_p} = \frac{1}{2}v_F^2 \int_0^\infty e^{-t/\tau_p} dt = \frac{1}{2}v_F^2 \tau_p \tag{3.14}
$$

The conductance rather than the conductivity is usually measured in experiments. The conductivity σ relates the local current density to the electric field, $j = \sigma E$, while the conductance G relates the total current to the voltage drop, $I = GV$. Because the conductance for 2DEG system of width W and length L in the current direction is

$$
G = \frac{W}{L}\sigma \tag{3.15}
$$

So the conductance is identical to the conductivity in a large homogeneous conductor (squared sample) and is usually called squared conductance. If we disregard the effects of phase coherence, Eq. (3.15) is only correct when the sample sizes are much larger than mean free path ($W, L \gg \ell_e$). This is the diffusive transport regime, illustrated in Fig. 3.1a. When the dimensions of the sample are reduced below the mean free path, one enters the ballistic transport regime, shown in Fig. 3.1c. One can further distinguish an intermediate quasi-ballistic regime, characterized by $W < l < L$ (see

Fig. 3.1 Three transport regimes (Figure is taken from [1])

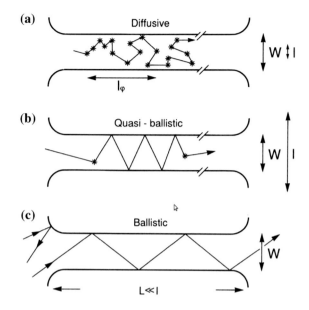

Fig. 3.1b). Three these transport regimes will be discussed carefully in Sect. 3.1.4. In ballistic transport, only the conductance plays a role; not the conductivity because the conductivity diverges in this regime. The Landauer formula

$$G = \frac{e^2}{h}T \qquad (3.16)$$

plays a central role in the study of ballistic transport because it expresses the conductance in terms of a Fermi level property of the sample (the transmission probability T). Equation (3.16) can therefore be applied to situations where the conductivity does not exist as a local quantity.

At a finite temperature T, a chemical potential (or Fermi energy) gradient ∇E_F induces a diffusion current that is smeared out over an energy range of order $k_B T$ around E_F. The energy interval between E and $E + dE$ contributes an amount $d\mathbf{j}$ to the diffusion current density $\mathbf{j}$, given by

$$d\mathbf{j}_{diff} = -D\nabla \{\rho(E)f(E - E_F)dE\} = -dE D\rho(E)\frac{df}{dE_F}\nabla E_F \qquad (3.17)$$

where D is the diffusion constant taken at energy E. The integration over E gives the total diffusion current density:

$$\mathbf{j} = -\nabla E_F e^{-2} \int_0^\infty dE \sigma(E, 0)\frac{df}{dE_F} \qquad (3.18)$$

where $\sigma(E, 0)$ is the zero temperature conductivity in Eq. (3.11) for a Fermi energy equal to E. The requirement of vanishing current for a spatially constant electrochemical potential implies that the conductivity $\sigma(E_F, T)$ at temperature T and Fermi energy E_F satisfies

$$\sigma(E_F, T)e^{-2}\nabla E_F + \mathbf{j} = 0 \tag{3.19}$$

Therefore, the finite-temperature conductivity is given simply by the energy average of the zero-temperature result [1]

$$\sigma(E_F, T) = \int_0^\infty dE \sigma(E, 0) \frac{df}{dE_F} \tag{3.20}$$

As $T \to 0$, $df/dE_F \to \delta(E - E_F)$, so indeed only $E = E_F$ contributes to the energy average. Result (3.20) contains exclusively the effects of a finite temperature that are due to the thermal smearing of the Fermi-Dirac distribution. A possible temperature dependence of the scattering processes is not taken into account.

3.1.3 The Kubo-Greenwood Formula

Hereafter, we will briefly derive the Kubo-Greenwood formula for DC conductivity which will be applied to the real-space calculation to compute the conductivity, mean free path, mobility...etc. A more detailed derivation can be found in Ref. [5].

Let's consider an electron in an weak electric field pointing along the x direction

$$\mathbf{E} = E_0 \cos(\omega t) u_x \tag{3.21}$$

The weak vector potential of this field $A(t) = -\frac{E_0}{2i\omega}(e^{i\omega t} - e^{-i\omega t})u_x$ will add a perturbative term $\delta H(t) = \frac{2e\hat{P}A(t)}{2m} = -\frac{eE_0}{2i\omega}(e^{i\omega t} - e^{-i\omega t})v_x$ to electron Hamiltonian H_0 which includes any interactions in the absence of electric field.

In perturbation theory, the total wavefunction $\Psi_m(t)$ in the presence of electric field can be expanded in terms of nonperturbed wavefunctions ψ_n: $\Psi_m(t) = \sum_n a_n(t)\psi_n$ where the coefficients $a_n(t)$ can be approximately determined by first-order perturbation theory. The transition from initial states $|n\rangle$ at $t = 0$ to final states $|m\rangle$ at t in this theory is given by

$$p_{nm}(t) = \frac{1}{\hbar^2}\left|\int_0^t d\tau e^{i(E_m - E_n)\tau/\hbar}\langle m|\delta H(\tau)|n\rangle\right|^2 \tag{3.22}$$

And the transition rate at long time

$$\frac{p_{nm}}{t} = \frac{2\pi}{\hbar}\left(\frac{eE_0}{2\omega}\right)^2 |\langle m|v_x|n\rangle|^2 [\delta(E_m - (E_n + \hbar\omega)) + \delta(E_m - (E_n - \hbar\omega))] \tag{3.23}$$

The first term corresponds to the absorbed transition where electron in the initial state $|n\rangle$ with energy E_n absorbs an energy $\hbar\omega$ and changes to final state $|m\rangle$. Similarly, the second term is emitted transition. Hence the net rate of absorption of energy (total absorbed power) is given by

$$P = \frac{\pi e^2 E_0^2}{2\omega} \sum_{m,n} |\langle m|v_x|n\rangle|^2 \left[\delta\left(E_m - (E_n + \hbar\omega)\right) - \delta(E_m - (E_n - \hbar\omega)) \right]$$

(3.24)

This is the absorbed power for an isolated single electron. When we consider an electron ensemble, we have to take into account the Fermi distribution and the Pauli exclusion principle and 2 spin components.

$$P = \frac{\pi e^2 E_0^2}{\omega} \sum_{m,n} |\langle m|v_x|n\rangle|^2 \left(f(E_n)(1 - f(E_m)) \right) \left[\delta\left(E_m - (E_n + \hbar\omega)\right) - \delta(E_m - (E_n - \hbar\omega)) \right]$$

(3.25)

After some calculation and index permutation, we obtain

$$P = \pi \hbar e^2 E_0^2 \sum_{m,n} \frac{f(E_n) - f(E_m)}{\hbar\omega} |\langle m|v_x|n\rangle|^2 \, \delta\left(E_m - (E_n + \hbar\omega)\right) \qquad (3.26)$$

On the other hand, the total absorbed is defined as $P = \frac{1}{2}\sigma(\omega)\Omega E_0^2$, where $\sigma(\omega)$ is the conductivity of the system. By substituting this formula into the above equation we obtain

$$\sigma(\omega) = \frac{2\pi e^2 \hbar}{\Omega} \sum_{m,n} \frac{f(E_n) - f(E_m)}{\hbar\omega} |\langle m|v_x|n\rangle|^2 \, \delta\left(E_m - (E_n + \hbar\omega)\right) \qquad (3.27)$$

This expression can be rewriten in below integral

$$\sigma(\omega) = \frac{2\pi e^2 \hbar}{\Omega} \int_{-\infty}^{\infty} dE \frac{f(E) - f(E + \hbar\omega)}{\hbar\omega} \sum_{m,n} |\langle m|v_x|n\rangle|^2 \, \delta\left(E + \hbar\omega - E_m\right) \delta\left(E - E_n\right)$$

(3.28)

By inserting delta function into the *braket* we have

$$\sigma(\omega) = \frac{2\pi e^2 \hbar}{\Omega} \int_{-\infty}^{\infty} dE \frac{f(E) - f(E + \hbar\omega)}{\hbar\omega} \mathrm{Tr}\left[v_x \delta(E - \hat{H}) v_x \delta(E + \hbar\omega - \hat{H}) \right]$$

(3.29)

We use the Fermi-Dirac distribution function property $\lim_{\omega \to 0} \frac{f(E) - f(E + \hbar\omega)}{\hbar\omega} = \delta(E - E_F)$. In the limit of zero temperature the DC conductivity is

$$\sigma_{DC}(E) = \frac{2\pi e^2 \hbar}{\Omega} \mathrm{Tr}\left[v_x \delta(E - \hat{H}) v_x \delta(E - \hat{H}) \right] \qquad (3.30)$$

where Ω is the volume of the system. The last delta-function is rewritten as an integral

$$\delta(E - \hat{H}) = \frac{1}{2\pi\hbar} \int_{-\infty}^{\infty} dt e^{i(E-\hat{H})t/\hbar} \qquad (3.31)$$

and inserted into Eq. (3.30):

$$\sigma_{DC}(E) = \frac{e^2}{\Omega} \int_{-\infty}^{\infty} dt \mathrm{Tr} \left[v_x \delta(E - \hat{H}) e^{iEt/\hbar} v_x e^{-i\hat{H}t/\hbar} \right]$$

$$= \frac{e^2}{\Omega} \int_{-\infty}^{\infty} dt \mathrm{Tr} \left[v_x(0)\delta(E - \hat{H})v_x(t) \right] \qquad (3.32)$$

where the final formula is obtained thanks to the delta function $\delta(E - \hat{H})$ and $v_x(t) = e^{i\hat{H}t/\hbar} v_x e^{-i\hat{H}t/\hbar}$ is velocity operator in Heisengberg representation.

Now we will show that the formula for the quantum average of any operator $\hat{Q}$ at a given energy E is written as

$$\langle \hat{Q} \rangle_E = \frac{1}{N} \sum_i^N \langle \psi_E^i | \hat{Q} | \psi_E^i \rangle = \frac{\mathrm{Tr} \left[\delta(E - \hat{H})\hat{Q} \right]}{\mathrm{Tr} \left[\delta(E - \hat{H}) \right]} \qquad (3.33)$$

where $|\psi_E^i\rangle$ are N degenerate eigenstates of $\hat{H}$, all having energy E.

Indeed,

$$\frac{\mathrm{Tr} \left[\delta(E - \hat{H})\hat{Q} \right]}{\mathrm{Tr} \left[\delta(E - \hat{H}) \right]} = \frac{\sum_{i,n} \langle \psi_{E_n}^i | \delta(E - \hat{H})\hat{Q} | \psi_{E_n}^i \rangle}{\sum_{i,n} \langle \psi_{E_n}^i | \delta(E - \hat{H}) | \psi_{E_n}^i \rangle}$$

$$= \frac{\sum_{i,n} \langle \psi_{E_n}^i | \delta(E - E_n)\hat{Q} | \psi_{E_n}^i \rangle}{\sum_{i,n} \langle \psi_{E_n}^i | \delta(E - E_n) | \psi_{E_n}^i \rangle}$$

$$= \frac{1}{N} \sum_i^N \langle \psi_E^i | \hat{Q} | \psi_E^i \rangle$$

Using Eq. (3.33) one can rewrite Eq. (3.32) in form

$$\sigma_{DC}(E) = e^2 \frac{\mathrm{Tr} \left[\delta(E - \hat{H}) \right]}{\Omega} \int_{-\infty}^{\infty} dt \langle v_x(t)v_x(0) \rangle_E$$

$$= e^2 \rho(E) \int_0^{\infty} dt \left(\langle v_x(t)v_x(0) \rangle_E + \langle v_x(-t)v_x(0) \rangle_E \right)$$

$$= e^2 \rho(E) \int_0^{\infty} dt \langle v_x(t)v_x(0) + v_x(0)v_x(t) \rangle_E$$

$$= e^2 \rho(E) \int_0^\infty dt\, C(E, t) \tag{3.34}$$

with the total density of state $\rho(E) = \dfrac{\text{Tr}\left[\delta(E-\hat{H})\right]}{\Omega}$ and the velocity autocorrelation function $C(E, t) = \langle\{v_x(t), v_x(0)\}\rangle_E$.

Now, let's define the mean value of the spreading in the x-direction of states having energy E

$$\Delta X^2(E, t) = \langle|\hat{X}(t) - \hat{X}(0)|^2\rangle_E \tag{3.35}$$

The time derivative of the spreading gives

$$\frac{d}{dt}\Delta X^2(E, t) = \langle v_x(t)(\hat{X}(t) - \hat{X}(0)) + (\hat{X}(t) - \hat{X}(0))v_x(t)\rangle_E \tag{3.36}$$

Changing the time arguments and then taking the derivative allow us to write the second derivative of $\Delta X^2(E, t)$ as

$$\frac{d}{dt}\Delta X^2(E, t) = \langle v_x(0)v_x(-t) + v_x(-t)v_x(0)\rangle_E \tag{3.37}$$

Changing the time arguments again we finally get

$$\frac{d^2}{dt^2}\Delta X^2(E, t) = C(E, t) \tag{3.38}$$

Replacing above equation into Eq. (3.34) we get the final expression for conductivity which is usually used in the calculation

$$\sigma_{DC}^1(E) = e^2 \rho(E) \lim_{t\to\infty} \frac{d}{dt}\Delta X^2(E, t) \tag{3.39}$$

with the spreading of the wavepacket in the x-direction

$$\Delta X^2(E, t) = \frac{\text{Tr}\left[\delta(E - \hat{H})\left(\hat{X}(t) - \hat{X}(0)\right)^2\right]}{\text{Tr}\left[\delta(E - \hat{H})\right]} \tag{3.40}$$

An alternative definition, in which the derivative in the Eq. (3.39) is replaced by a division

$$\sigma_{DC}^2(E) = e^2 \rho(E)\frac{\Delta X^2(E, t)}{t} = e^2 \rho(E) D_x(E, t) \tag{3.41}$$

is frequently used, because it gives smoother curves for the conductivity than Eq. (3.39) does.

The similarity of Eqs. (3.41) and (3.11) shows that the diffusion coefficient is given by

$$D_x(E, t) = \frac{\Delta X^2(E, t)}{t} \tag{3.42}$$

3.1.4 Three Transport Regimes

As mentioned in Sect. 3.1.2 with the semiclassical approach, the motion of an electron in the sample can be in three different regimes depending on sample size and disorders. In this section we will discuss in more detail about three these regimes based on the diffusion coefficient $D_x(E, t)$ in above quantum approach.

Ballistic regime

In the absence of disorder, structural imperfections or the distance between the source and drain L is much smaller than mean free path ($L \ll \ell_e$), electron propagates in ballistic regime. In this regime electron moves at a constant velocity v_F so the spreading of wave packet $\Delta X(t)$ will increase linearly with time t with the slope of v_F, $\Delta X(t) = v_F t$. Therefore, the form of the diffusion coefficient $D_x(E_F, t)$ in expression (3.42) is a line with slope v_F^2, $D_x(E_F, t) = v_F^2 t$ (See Fig. 3.2a). These lead to the divergence of the long-time conductivity $\sigma_{DC}(E_F)$ in Eq. (3.39) (Equation 3.41 is only correct in the diffusive regime)

$$\sigma_{DC}(E_F)_{bal.} = e^2 \rho(E_F) \lim_{t \mapsto \infty} \frac{d}{dt} \Delta X^2(E_F, t) = 2e^2 \rho(E_F) \lim_{t \mapsto \infty} v_F^2 t \tag{3.43}$$

As pointed out in Ref. [5], this divergence is due to the fact that when deriving the linear response theory, a finite dissipation source, intrinsic to the sample, is

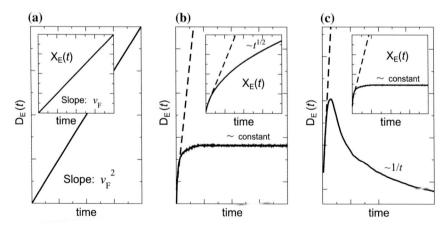

Fig. 3.2 Diffusion coefficient (*main frame*) and displacement (*inset*) in three transport regimes: **a** Ballistic regime, **b** diffusive regime and **c** localized regime

introduced both physically and mathematically. The ballistic limit is therefore not well defined in this formalism. However, we can find the agreement between this method and Landauer-Buttiker formalism in one dimensional (1D) sample. Indeed, the formula for the conductance in Eq. (3.15) for 1D is $G = \sigma/L$ where $L = 2v_F t$ is the propagating length of wave packet. Using these formulas and Eq. (3.43) we get

$$G(E_F) = 2e^2 \rho_{1D}(E_F) \lim_{t \to \infty} \frac{v_F^2 t}{L} = e^2 \rho_{1D}(E_F) v_F = \frac{2e^2}{h} = G_0 \qquad (3.44)$$

where the density of states for 1D is $\rho_{1D}(E_F) = 1/\pi \hbar v_F$ and G_0 is the conductance quantum corresponding to the conductance of one conducting channel. If there are more than one conducting channel crossing the Fermi level, the conductance is proportional to the number of conducting channels, as expected from the Landauer formula.

Diffusive regime

As mentioned in Sect. 3.1.2 in semiclassical approach, the diffusion coefficient converges in diffusive regime to a constant which is proportional to elastic scattering time (See Eq. 3.12), $D_{diff}^{sc} = \frac{1}{2}v^2 \tau_p$. We can obtain a similar formula for the Kubo approach by considering the Eq. (3.38) with the velocity autocorrelation function $C(E, t) = 2v_x^2(E)e^{-t/\tau_p}$ in diffusive regime. This leads to the $\sqrt{t}$ behaviour of displacement $\Delta X(E, t)$ and a constant diffusion coefficient (See Fig. 3.2b).

$$\lim_{t \to \infty} \Delta X^2(E, t) = 2\tau_p(E)v_x^2(E)t \quad \text{and} \quad \lim_{t \to \infty} D_x(E, t) \longmapsto 2\tau_p(E)v_x^2(E)$$
$$(3.45)$$

The Kubo formula for diffusive regime then gives access to the semiclassical conductivity (σ_{sc})

$$\sigma_{sc}(E) = \sigma_{DC}(E)_{diff} = e^2 \rho(E) D_x(E, t)$$
$$\sigma_{sc}(E) = 2e^2 \rho(E) \tau_p(E) v_x^2(E) \qquad (3.46)$$
$$\sigma_{sc}(E) = 2e^2 \rho(E) v_x(E) \ell_e^x(E)$$

with the projection of mean free path in the x direction $\ell_e^x(E) = \tau_p(E).v_x(E)$.

Localized regime

If electron propagates in the system with strong disorder, the back scattering will give rise to quantum interference which leads electron into the localized regime in which the diffusion coefficient decrease roughly following $\sim 1/t$. Because electron is strongly localized, the spreading $\Delta X(E, t)$ reaches an asymptotic value that is related to the localization length $\xi(E)$ (See Fig. 3.2c).

The propagation of a wave packet of electrons in the real disorder system usually experiences three above regimes. Assuming that we put a wave packet locally in

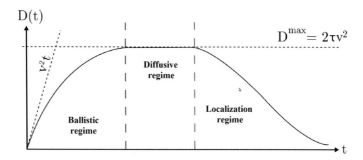

Fig. 3.3 Illustration of the time dependence of diffusion coefficient $D(E, t)$

the sample, at first electron moves very fast in the ballistic regime with velocity $v_F = \sqrt{v_x^2 + v_y^2}$ and diffusion coefficient

$$D(E, t) = D_x(E, t) + D_y(E, t) = v_F^2 t \qquad (3.47)$$

After exploring the sample, the disorder drives electron into the diffusive regime with the saturation of diffusion coefficient at the value D^{max} (See Fig. 3.3) which can be used to estimate value of elastic scattering time τ_p and mean free path ℓ_e. Indeed, using Eq. (3.45) we can extract (assuming the sample is isotropic)

$$\ell_e = v_F \tau_p = \frac{D^{max}}{2 v_F} \qquad (3.48)$$

which leads us to following formulas for σ_{sc} (See Eq. 3.46)

$$
\begin{aligned}
\sigma_{sc}(E) &= \sigma_{xx} = \frac{1}{2} e^2 \rho_0(E) D_x^{max} = \frac{1}{4} e^2 \rho_0(E) D^{max} \\
\sigma_{sc}(E) &= \frac{1}{2} e^2 \rho_0(E) v_F^2(E) \tau_p(E) \\
\sigma_{sc}(E) &= \frac{1}{2} e^2 \rho_0(E) v_F(E) \ell_e(E)
\end{aligned}
\qquad (3.49)
$$

with $\rho_0(E) = 2\rho(E)$ is the total density of state (the factor of 2 accounts for spin degeneracy) which is given by Eq. (2.17) for perfect graphene. Electron mobility μ in this regime is given by

$$\mu(E) = \frac{\sigma_{sc}(E)}{e n(E)} \qquad (3.50)$$

where $n(E) = \int \rho_0(E) dE$ is the charge density.

After the diffusive regime, depending on the disorder source, an electron might enter the localization regime due to quantum interference effects. The electrons start to localize resulting in a decreasing diffusion coefficient. The quantum interference gives the quantum correction $\delta\sigma(L)$ to the semiclassical conductivity [5]

$$\sigma(L) = \sigma_{sc} + \delta\sigma(L) \quad \text{with} \quad \delta\sigma(L) = -\frac{2e^2}{\pi h} ln\left(\frac{L}{\ell_e}\right) \tag{3.51}$$

The transition to the insulating state is continuous and reached when the quantum correction is of the same order of the semiclassical conductivity, that is when $\Delta\sigma(L = \xi) \simeq \sigma_{sc}$. This let us to extract the localization length ξ

$$\xi = \ell_e \exp(\pi\sigma_{sc}/G_0) \tag{3.52}$$

3.1.5 The Kubo Formalism in Real Space

In this section, I will present an efficient real space implementation of the Kubo formula which is mainly used in this thesis to study quantum transport in graphene. This method was first developed by Roche and Mayou in 1997 for the study of quasiperiodic systems [6] and was then adapted by Stephan and coworkers to study electric transport in disordered mesoscopic systems. The advantage of this method is that the quantum transport of large systems can be investigated thanks to linear-scaling time consuming. The maximum number of orbitals in the system can be studied at the moment is hundred millions and the simulation of samples with 1 billion orbitals can be envisioned in the next decade.

There are two major problems one needs to overcome when using the Eq. (3.41) to calculate the conductivity. The first one is how to change the Eq. (3.40) to the simpler expression. The second one is how to calculate the trace without finding the eigenstates of the system.

Let's find the solution for the first problem by rearranging terms in the expression (3.40) with the cyclic property of trace.

$$\Delta X^2(E, t) = \frac{\text{Tr}\left[\delta(E - \hat{H})|\hat{X}(t) - \hat{X}(0)|^2\right]}{\text{Tr}\left[\delta(E - \hat{H})\right]} \tag{3.53}$$

$$\Delta X^2(E, t) = \frac{\text{Tr}\left[(\hat{X}(t) - \hat{X}(0))^{\dagger}\delta(E - \hat{H})(\hat{X}(t) - \hat{X}(0))\right]}{\text{Tr}\left[\delta(E - \hat{H})\right]} \tag{3.54}$$

We then use several identities and definitions to rewrite $(\hat{X}(t) - \hat{X}(0))$

$$\hat{X}(t) = e^{\frac{i\hat{H}t}{\hbar}} \hat{X}(0) e^{\frac{-i\hat{H}t}{\hbar}} \tag{3.55}$$

$$\hat{U}(t) = e^{\frac{-i\hat{H}t}{\hbar}} \tag{3.56}$$

where $\hat{U}(t)$ is the evolution operator,

$$\hat{X}(t) - \hat{X}(0) = \hat{U}^{\dagger}(t)\hat{X}\hat{U}(t) - \hat{X} \tag{3.57}$$

$$\hat{X}(t) - \hat{X}(0) = \hat{U}^{\dagger}(t)\hat{X}\hat{U}(t) - \hat{U}^{\dagger}(t)\hat{U}(t)\hat{X} \tag{3.58}$$

$$\hat{X}(t) - \hat{X}(0) = \hat{U}^{\dagger}(t)[\hat{X}, \hat{U}(t)] \tag{3.59}$$

using $\hat{U}^{\dagger}(t)\hat{U}(t) = \mathbb{I}$, and $[\cdots, \cdots]$ the commutator. Then by replacing these quantities in Eq. (3.54), one gets

$$\Delta X^2(E, t) = \frac{\mathrm{Tr}\left[[\hat{X}, \hat{U}(t)]^{\dagger}\hat{U}(t)\delta(E - \hat{H})\hat{U}^{\dagger}(t)[\hat{X}, \hat{U}(t)]\right]}{\mathrm{Tr}\left[\delta(E - \hat{H})\right]} \tag{3.60}$$

$$\Delta X^2(E, t) = \frac{\mathrm{Tr}\left[[\hat{X}, \hat{U}(t)]^{\dagger}\delta(E - \hat{H})[\hat{X}, \hat{U}(t)]\right]}{\mathrm{Tr}\left[\delta(E - \hat{H})\right]} \tag{3.61}$$

This is the simplest form of $\Delta X^2(E, t)$ that we use for the numerical calculation.

Now the next mission is how to evaluate the trace without calculating the eigenstates of the system which costs a lot of time. The way that we do it is approximating the trace by expectation values on random phase states which are expanded on all the orbitals $|i\rangle$ of the basis set

$$|\varphi_{RP}\rangle = \frac{1}{\sqrt{N}} \sum_{i=1}^{N} e^{2i\pi\alpha_i} |i\rangle, \tag{3.62}$$

where α_i is a random number in the $[0, 1]$ range. An average over few tens of random phases states is usually sufficient to calculate the expectation values. Using this stratergy for the expression (3.61), we find

$$\Delta X^2(E, t) = \frac{\langle\varphi_{RP}|[\hat{X}, \hat{U}(t)]^{\dagger}\delta(E - \hat{H})[\hat{X}, \hat{U}(t)]|\varphi_{RP}\rangle}{\langle\varphi_{RP}|\delta(E - \hat{H})|\varphi_{RP}\rangle} \tag{3.63}$$

$$\Delta X^2(E, t) = \frac{\langle\varphi'_{RP}(t)|\delta(E - \hat{H})|\varphi'_{RP}(t)\rangle}{\langle\varphi_{RP}|\delta(E - \hat{H})|\varphi_{RP}\rangle} \tag{3.64}$$

The numerator and the denominator have the same form. The techniques used for the computation of the density of states can thus be also employed for the computation of $\Delta X^2(E, t)$ provided one first evaluates $|\varphi'_{RP}(t)\rangle = [\hat{X}, \hat{U}(t)]|\varphi_{RP}\rangle$. The evaluation of $|\varphi'_{RP}(t)\rangle$ needs $[\hat{X}, \hat{H}]$ together with $\hat{U}(t)|\varphi_{RP}\rangle$ which can be done by expanding $\hat{U}(t)$ into the Chebyshev polynomials. The detail techniques are mentioned in Appendix A. Moreover, the density of state is calculated by the Lanczos method [7,

Fig. 3.4 The application of
Kubo formalism in real
space: Velocity and density
of states for pristine
graphene

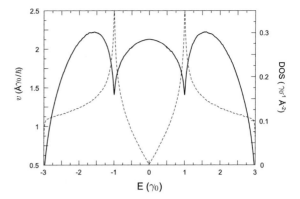

8] in which we tridiagonalize the Hamiltonian and then calculate the density of state
in form of continued fraction. The detailed techniques are in Appendix B.

Figure 3.4 shows the velocity and density of states for pristine graphene obtained
by the application of the real space method. The value of velocity close to the Dirac
point is $v_F = 2.13(\text{Å}\gamma_0\hbar^{-1})$ which is the same value as extracted from the band
structure $v_F = \sqrt{3}\gamma_0 a/2\hbar$. The density of states shows the linear behaviour in the
vicinity of Dirac point which coincides with the Eq. (2.17) in Chap. 2. Furthermore,
the energy dependence of the density of states has been confirmed by other calculations [9] which considered only the nearest neighbor hopping term in the Hamiltonian. These evidences validate the application of this method for the electronic
transport calculation of mesoscopic system.

Figure 3.5 visualizes the propagation of wave packet in the real space of Poly-G
in which its small portion is shown in Fig. 3.5a. Figure 3.5b–d shows some snapshots
of the time evolution of a wave packet within a Poly-G sample, highlighting the
scattering and localizing effects around the GGBs. Indeed, the wave packet is initially
injected on a hexagon at a grain center and begins to propagate in ballistic regime in
Fig. 3.5b. When electrons meet the GBs (Fig. 3.5c), the scattering happens because
of the structural disorders on GBs and the misorientation of grains. These scatterings
drive electrons into the diffusive regime and finally to the localization regime as
shown in Fig. 3.5d.

3.2 Spin Transport Formalism

Spintronics is an interesting branch of electronics in which the electron spin is
exploited and manipulated to apply it to quantum information processing, quantum
computation, etc. The research of spintronics has increased a lot since the discovery
of giant magnetoresistance by Albert Fert et al. [10] and Peter Grunberg et al. [11]
and especially after the theoretical proposal of a spin field-effect-transistor by Datta
and Das [12] in 1990.

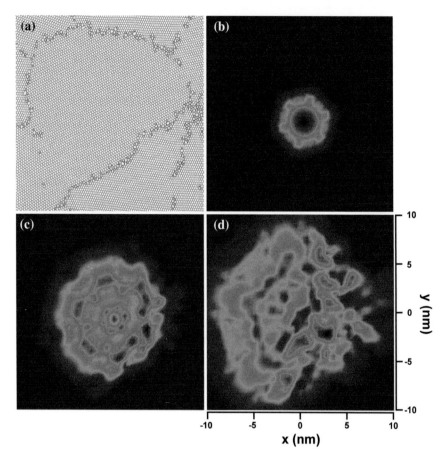

Fig. 3.5 The visualization of real space method in Poly-G. **a** Small portion of a Poly-G sample. **b–d** Time evolution of a wave packet within the sample

Graphene is a good candidate for spintronics due to low SOC and hyperfine interaction, but the agreement between theoretical and experimental results is still missing at the moment. Up to now, most of dynamical characteristics of spin is extracted from the kinetic spin Bloch equation. Here, we first develop a new method to study the spin dynamics of mesoscopic systems and use it to address the controversial topic of spin relaxation in graphene.

3.2.1 Wavefunction and Random Phase State with Spin

In order to include spin in the wavefunction we use the two-component spinor to represent the spin-wavefunction

Fig. 3.6 Spherical
coordinate system for spin

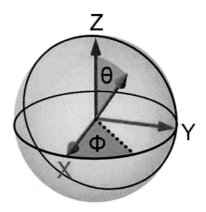

$$|\Psi\rangle = \begin{pmatrix} \Psi_\uparrow \\ \Psi_\downarrow \end{pmatrix} \tag{3.65}$$

And the random-phase state corresponding to Eq. (3.62) is

$$|\Psi_{RP}\rangle = \frac{1}{\sqrt{N}} \sum_{i=1}^{N} \begin{pmatrix} \cos\left(\frac{\theta_i}{2}\right) \\ e^{i\Phi_i} \sin\left(\frac{\theta_i}{2}\right) \end{pmatrix} e^{2i\pi\alpha_i} |i\rangle, \tag{3.66}$$

where (Φ_i, θ_i) is the spin orientation of electron of orbital $|i\rangle$ in spin spherical coordinate system (Fig. 3.6).

3.2.2 Spin Polarization

The spin dynamics of the system is directly related to the time-dependence of spin polarization $S(t)$ which can be given by the expectation value of the spin Pauli operator.

$$S(t) = \langle \sigma(t) \rangle = \langle \psi(0)|\sigma(t)|\psi(0)\rangle \tag{3.67}$$

where $\sigma(t) = e^{\frac{i\hat{H}t}{\hbar}} \sigma e^{\frac{-i\hat{H}t}{\hbar}}$ is the spin operator in Heisengberg representation. However, this expectation gives the spin polarization for the whole spectrum which is not meaningful. Finding the expectation at specific energy is more important. In order to do so, we use the formula for quantum average of any operator at a given energy in Eq. (3.33).

$$S(E,t) = \frac{\text{Tr}\left[\delta(E-\hat{H})\sigma(t)\right]}{\text{Tr}\left[\delta(E-\hat{H})\right]} = \frac{\text{Tr}\left[\delta(E-\hat{H})\sigma(t) + \sigma(t)\delta(E-\hat{H})\right]}{2\text{Tr}\left[\delta(E-\hat{H})\right]} \tag{3.68}$$

Approximating the trace by expectation values on random phase states $|\psi(0)\rangle = |\varphi_{RP}\rangle$ is the strategy to get a faster calculation.

$$S(E,t) = \frac{\langle\psi(0)|\delta(E-\hat{H})\sigma(t) + \sigma(t)\delta(E-\hat{H})|\psi(0)\rangle}{2\langle\psi(0)|\delta(E-\hat{H})|\psi(0)\rangle} \quad (3.69)$$

$$= \frac{\langle\psi(t)|\delta(E-\hat{H})\sigma + \sigma\delta(E-\hat{H})|\psi(t)\rangle}{2\langle\psi(0)|\delta(E-\hat{H})|\psi(0)\rangle} \quad (3.70)$$

where the time evolution of the wavepackets $|\psi(t)\rangle = e^{\frac{-i\hat{H}t}{\hbar}}|\psi(0)\rangle$ is obtained by solving the time-dependent Schrödinger equation. This is the equation we use for the calculation of spin polarization.

Let's denote the quantity in the numerator of Eq. (3.70) as

$$\mathbf{P}(E,t) = \langle\psi(t)|\sigma\delta(E-\hat{H})|\psi(t)\rangle \quad (3.71)$$

Equation (3.70) becomes

$$S(E,t) = \frac{\mathfrak{Re}\,(\mathbf{P}(E,t))}{\langle\psi(0)|\delta(E-\hat{H})|\psi(0)\rangle} \quad (3.72)$$

The denominator is directly proportional to the density of states $\rho(E)$ and can be computed by the real space method given in Sect. 3.1.5 while the numerator can be calculated by including the energy resolution η

$$\mathbf{P}(E,t) = \langle\psi(t)|\sigma\delta(E-\hat{H})|\psi(t)\rangle$$

$$= \langle\psi(t)|\sigma\frac{1}{2\pi}\left[\frac{1}{\eta - i(E-\hat{H})} + \frac{1}{\eta + i(E-\hat{H})}\right]|\psi(t)\rangle$$

$$= \frac{1}{2\pi}\sum_j \langle\psi(t)|\sigma|\phi_j\rangle\langle\phi_j|\left[\frac{1}{\eta - i(E-\hat{H})} + \frac{1}{\eta + i(E-\hat{H})}\right]|\psi(t)\rangle$$

$$= \frac{i}{2\pi}\sum_j \mu_j\langle\phi_j|\left[\frac{1}{E+i\eta-\hat{H}} - \frac{1}{E-i\eta-\hat{H}}\right]|\psi(t)\rangle$$

$$\mathbf{P}(E,t) = \frac{i}{2\pi}\sum_j \mu_j\left[\langle\phi_j|\frac{1}{z-\hat{H}}|\psi(t)\rangle - \langle\phi_j|\frac{1}{z^*-\hat{H}}|\psi(t)\rangle\right]$$

where $\mu_j = \langle\psi(t)|\sigma|\phi_j\rangle$ with any complete basic set $\{|\phi_j\rangle\}$ and $z = E+i\eta$.

By building a orthonormal basis with the Lanczos method (See Appendix B) beginning with $|\phi_1\rangle = |\psi(t)\rangle$, we have

$$\mathbf{P}(E,t) - \frac{i}{2\pi}\sum_{j=1}\mu_j\left[\left(\frac{1}{z-H}\right)_{j,1} - \left(\frac{1}{z^*-H}\right)_{j,1}\right] \quad (3.73)$$

where H is the tridiagonal matrix of $\hat{H}$ in the Lanczos basis (See Appendix B)

$$H = \left(H_{ij}\right) = \begin{pmatrix} a_1 & b_1 & 0 & \cdots \\ b_1 & a_2 & b_2 & \\ 0 & b_2 & a_3 & \cdots \\ \vdots & & \vdots & \ddots \end{pmatrix} \tag{3.74}$$

3.2.3 Technical Details

Now what we need is the first column of the inverted matrices $z - H$ and $z^* - H$ which we call κ and $\bar{\kappa}$, respectively

$$(z - H)K = 1 \quad \Rightarrow \quad \sum_n (z - H)_{mn}\kappa_n = \delta_{m1} \tag{3.75}$$

writing above equation explicitly

$$(z - H_{11})\kappa_1 - H_{12}\kappa_2 = 1$$
$$-H_{21}\kappa_1 + (z - H_{22})\kappa_2 - H_{23}\kappa_3 = 0$$
$$\vdots$$
$$-H_{n,n-1}\kappa_{n-1} + (z - H_{nn})\kappa_n - H_{n,n+1}\kappa_{n+1} = 0$$

From κ_1 we can get the others

$$\kappa_2 = \frac{(z - H_{11})\kappa_1 - 1}{H_{12}}$$
$$\kappa_3 = \frac{(z - H_{22})\kappa_2 - H_{21}\kappa_1}{H_{23}}$$
$$\vdots$$
$$\kappa_n = \frac{(z - H_{n-1,n-1})\kappa_{n-1} - H_{n-1,n-2}\kappa_{n-2}}{H_{n-1,n}}$$

We can do the same for $\bar{\kappa}$ just by replacing z by z^*. Using the fact that $\bar{\kappa}_1 = \kappa_1^*$, we can show that $\bar{\kappa}_j = \kappa_j^*$. Finally, we get the formula for Eq. (3.73)

$$\mathbf{P}(E, t) = \frac{i}{2\pi} \sum_{j=1} \mu_j \left[\kappa_j - \bar{\kappa}_j\right]$$
$$\mathbf{P}(E, t) = -\frac{1}{\pi} \sum_{j=1} \mu_j \Im\left(\kappa_j\right) \tag{3.76}$$

Substituting this formula into Eq. (3.72) leads to the final expression for spin polarization

$$S(E, t) = -\frac{1}{\pi \Omega \rho(E)} \sum_{j=1} \Re \left(\boldsymbol{\mu}_j \right) \Im \left(\kappa_j \right) \tag{3.77}$$

References

1. C.W.J. Beenakker, H. van Houten, Solid State Phys. **44**, 1–228 (1991)
2. N.W. Ashcroft, N.D. Mermin, *Solid State Physics* (New York, 1976)
3. S.D. Sarma, S. Adam, E.H. Hwang, E. Rossi, Rev. Mod. Phys. **83**, 407 (2011)
4. M. Toda, R. Kubo, N. Hashitsume, *Statistical Physics II* (Springer, Berlin, 1985)
5. E.F. Luis, F. Torres, S. Roche, J.C. Charlier, *Introduction to Graphene-Based Nanomaterials From Electronic Structure to Quantum Transport* (Cambridge, 2013)
6. S. Roche, D. Mayou, Phys. Rev. Lett. **79**, 2518 (1997)
7. R. Haydock, V. Heine, M.J. Kelly, J. Phys. C: Solid State Phys. **5**, 2845 (1972)
8. R. Haydock, V. Heine, M.J. Kelly, J. Phys. C: Solid State Phys. **8**, 2591 (1975)
9. S. Yuan, H. De Raedt, M.I. Katsnelson, Phys. Rev. B **82**, 115448 (2010)
10. M.N. Baibich, J.M. Broto, A. Fert, F. Nguyen Van Dau, F. Petroff, P. Etienne, G. Creuzet, A. Friederich, J. Chazelas. Phys. Rev. Lett. **61**, 2472–2475 (1988)
11. G. Binasch, P. Grnberg, F. Saurenbach, W. Zinn, Phys. Rev. B. **39**, 4828 (1988)
12. S. Datta, B. Das, Appl. Phys. Lett. **56**, 665 (1990)

Chapter 4
Transport in Disordered Graphene

Ideal crystalline graphene has exotic properties such as remarkably low dimensionality, high mobility and mechanical strength, tunable carrier type and density, etc. However, as with most other materials, defects are unavoidable during the preparation of graphene and can play a key role in many observables, and particularly electronic properties. The purpose of this chapter is to discuss the transport properties of realistic graphene with the increasing of disorder, beginning from single defects (vacancies) to line defects in Poly-G and finally to amorphous graphene, a strongly topological disordered graphene.

4.1 Transport Properties of Graphene with Vacancies

4.1.1 Introduction

The electronic transport properties of graphene are known to be very peculiar with unprecedented manifestations of quantum phenomena as Klein tunneling [1, 2], WAL [3, 4], or anomalous QHE [5, 6], all driven by a π-Berry phase stemming from graphene sublattice symmetry and pseudospin degree of freedom [7–9]. These fascinating properties, yielding high charge mobility [10, 11], are robust as long as disorder preserves a long range character. The fundamental nature of transport precisely at the Dirac point is however currently a subject of fierce debate and controversies. Indeed, for graphene deposited on oxide substrates, the nature of low-energy transport physics (as its sensitivity to weak disorder) is masked by the formation of electron-hole puddles [9]. A remarkable experiment has however recently demonstrated the possibility to screen out these detrimental effects [12], providing access

© Springer International Publishing Switzerland 2016   55
D.V. Tuan, *Charge and Spin Transport in Disordered Graphene-Based*
Materials, Springer Theses, DOI 10.1007/978-3-319-25571-2_4

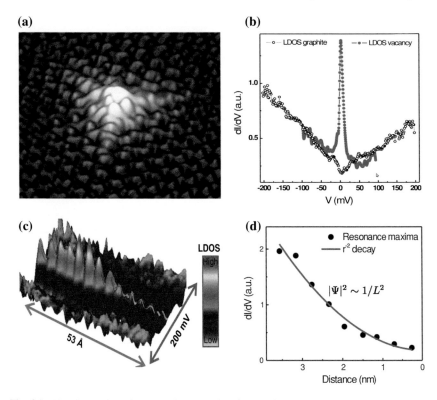

Fig. 4.1 The observation of ZEMs (Figure is taken from Ref. [15])

to the zero-energy Dirac physics. An unexpectedly large increase of the resistivity at the Dirac point was tentatively related to Anderson localization [12, 13] of unknown physical origin and questioned interpretation [14].

Of paramount importance are therefore the low-energy impurity states known as zero-energy modes (ZEMs) [16, 17], whose impact on the Dirac point transport physics needs to be clarified. ZEMs are predicted or observed for a variety of disorder classes, as topological defects (mainly vacancies) [17, 18], adatoms covalently bonded to carbon atoms [19, 20] and extended defects as GBs [21, 22]. As recently confirmed by scanning tunneling microscopy experiments on graphene monovacancies [15], ZEMs manifest as wave functions that decay as the inverse of the distance from the vacancy (See Fig. 4.1), exhibiting a puzzling quasi-localized character, whose consequences on quantum transport remain to date highly controversial. First, ZEMs have been predicted to produce a *supermetallic regime* by enhancing the Dirac-point conductivity above its minimum ballistic value $\sigma_{\min} = 4e^2/\pi h$ [23, 24], an unprecedented conducting state, which could be in principle explored experimentally [1, 25, 26]. Second, a similar increase of the Dirac point conductivity with defect density has been also reported in the diffusive regime of two-dimensional disordered graphene in the presence of vacancies or adatoms [20, 27]. These results contrast with the semiclassical conductivity found with the Boltzmann approach [18, 28–31], and suggest the absence of quantum interferences and localization effects observed for other types of disorder [32–34]. Finally, transport experiments in intentionally damaged graphene also report on puzzling conductivity fingerprints, whose physical origin remains to be fully understood [35, 36]. A comprehensive picture of the role of ZEMs on quantum transport properties in disordered graphene is therefore crucially missing and demands for further theoretical and experimental inspection.

This Section provides an extensive analysis of the contribution of zero-energy modes to quantum conduction close to the Dirac point in disordered graphene. Using Kubo-Greenwood and Landauer transport approaches, different regimes are numerically explored by changing the aspect ratio of the transport measurement geometry, and by tuning vacancy density and sublattice symmetry breaking features. The robustness of the supermetallic state induced by ZEMs is shown to be restricted to very low densities of compensated vacancies (equally distributed among both sublattices). This occurs as long as tunneling through evanescent modes prevails. In the absence of contact effects, an increase of the conductivity above $4e^2/\pi h$ is obtained for the semiclassical conductivity at the Dirac point and ascribed to a high density of ZEMs, but the quantum conductivity analysis unequivocally reveals a localization regime. For a totally uncompensated vacancy distribution (populating a single sublattice), the delocalization of ZEMs in real space is strongly prohibited for a large energy window around the Dirac point owing to the formation of a gap, whereas no appreciable difference of high energy transport (above the gap) is found compared with the compensated vacancy case. I would like to mention that some interesting cases of uncompensated impurities and defects have been reported experimentally [37–39], whose results demand further exploration.

4.1.2 Zero-Energy Modes and Transport Properties

System description and methodology. We consider a finite concentration n of vacancies either distributed at random exclusively on one of the two sublattices ($n_A = n$, the number of vacancies per carbon atoms in sublattice A and $n_B = 0$, uncompensated case), or equally distributed vacancies on both sublattices ($n_A = n_B = n/2$, compensated case). The electronic and transport properties are investigated by using a TB model with a single p_z orbital *per* atom and first nearest neighbor coupling. We model the vacancies by removing the corresponding orbitals from the Hamiltonian [16, 17]. To investigate the various transport regimes, two complementary approaches are used. To study two-dimensional (bulk) disordered graphene, real-space quantum wave packet dynamics and Kubo conductivity are calculated [32, 33, 40–43]. The zero-frequency conductivity $\sigma(E, t)$ for energy E and time t is given by Eq. (3.41). The diffusion coefficient $D(E, t)$ obtained by using Eq. (3.42) generally starts with a short-time ballistic motion followed by a saturation regime, which allows us to estimate the transport (elastic) mean free path ℓ_e and the semiclassical conductivity σ_{sc} from its maximum value as mentioned in Sect. 3.1.4. Depending on disorder strength, $D(E, t)$ is found to decay at longer times owing to quantum interferences, whose strength may dictate weak or strong (Anderson) localization at the considered time scale. Calculations are performed for systems containing several millions of carbon atoms, allowing the capture of all relevant transport regimes. We also study the ballistic limit of transport through finite graphene samples, by considering strip geometries with width W and length L (with $W/L \gg 1$) between two highly doped semi-infinite ribbons (of identical width). This two-terminal transport geometry gives access to the contribution of ZEMs in graphene transport when the charge flow is conveyed by contact-induced evanescent modes. The doping of contacts is simulated by adding an onsite energy of -1.5 eV to the corresponding orbitals, which generates a large DOS imbalance between the contacts and the central strip at the Dirac point ($E = 0$). The zero-temperature conductivity of the graphene strip is then computed as $\sigma(E) = (2e^2/h) \times T(E) \times L/W$, where $T(E)$ is the transmission coefficient evaluated within the Green's function approach [44, 45]. When $L \ll W$, low energy transport is dominated by tunneling through the undoped region yielding a universal ballistic value $\sigma(E \approx 0) \approx \sigma_{\min} = 4e^2/\pi h$ at the Dirac point for clean strips [1, 25, 26, 44].

ZEMs effects in two-dimensional disordered graphene. We start by considering the compensated case, which globally preserves the sublattice symmetry. Figure 4.2 (left inset) gives the density of states of the system as a function of the energy E

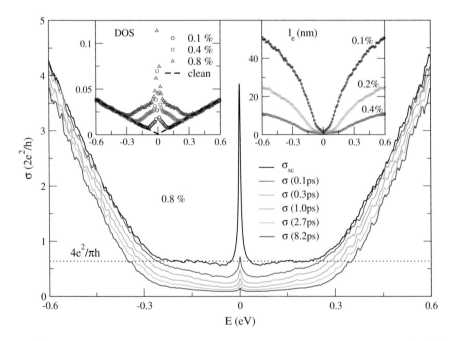

Fig. 4.2 *Main frame* Conductivity of graphene with $n = 0.8\%$ (compensated case): semiclassical value σ_{sc} (*solid line*), $\sigma_{\min} = 4e^2/\pi h$ (*dotted line*) and Kubo conductivity at various time scales. *Left inset* DOS for varying vacancy density, together with the pristine case (*dashed line*). *Right inset* Mean free paths for $n = 0.1; 0.2; 0.4\%$

for different vacancy densities n. In agreement with prior results [16, 17], the DOS shows the rise of a broad peak around $E = 0$, which witnesses the presence of ZEMs generated by disorder. Their nature however is not encoded in this feature but needs to be analyzed by studying transport characteristics such as mean free path (Fig. 4.2, right inset) and conductivity (Fig. 4.2, main frame). The mean free path ℓ_e is seen to be strongly energy dependent with minimum values close to the Dirac point, as expected for short-range scatterers [42, 43]. By increasing the vacancy density within the range [0.1 %, 0.4 %], ℓ_e drops from tens of nanometers down to few nanometers, and roughly varies as $\ell_e \sim 1/n$ in agreement with the Fermi golden rule. Interestingly, we find for the semiclassical conductivity $\sigma_{sc} \sim E$ for high enough energy (above 0.3 eV for $n = 0.8$ %), whereas it saturates to σ_{min} at low energy with a higher value around the Dirac point owing to the DOS enhancement induced by midgap states. When increasing the vacancy density, the minimum conductivity $4e^2/\pi h$ around the Dirac point extends over a larger energy region (not shown here).

The obtained short ℓ_e and minimum semiclassical conductivities suggest a strong contribution of quantum interferences, which is further evidenced by the decay of the Kubo conductivity below σ_{min} for sufficiently long time scales, see Fig. 4.2 (main frame). Depending on the energy, the observed downscaling of the quantum conductivity versus time can be described by a logarithmic correction (weak localization), an exponential decay (strong localization), or by localized modes beyond the Anderson regime. As shown in Fig. 4.3a, the quantum correction to the conductivity ($\delta\sigma(\lambda) = \sigma(\lambda) - \sigma_{sc}$) at $E = 0.4$ eV is numerically found to downscale as $\delta\sigma(\lambda) \sim -2e^2/(\pi h) \ln(\lambda/\lambda_e)$ (with $\lambda \equiv \sqrt{\Delta X^2(t)}$ the time-dependent wave packet space extension and λ_e related to ℓ_e [46]) and broadening-independent (a strongly reduced broadening $\eta = 0.8$ meV yields the same conductivity). Differently, for $E = 0.2$ eV (see Fig. 4.3b for different energy resolutions from $\eta = 3$ meV down to $\eta = 0.4$ meV) the length-dependent conductivity exhibits an exponential behavior $\sigma \sim \exp(-\lambda/\xi)$ (ξ the localization length), evidencing a strong-localization regime [13]. Note that this scaling law is observed independently of the energy precision parameter η, thus indicating that our approach is able to unambiguously catch the physics of the system and that there is only a residual quantitative, but not qualitative, dependence on η. Moreover, the localization length varies only weakly at lowest η indicating a limit $\xi \approx 10$ nm when $\eta \to 0$, which confirms the reliability of the numerical simulation. Exactly at the Dirac point, localization is observed in Fig. 4.3c since the conductivity decays with length λ. However, in contrast to finite energies, it follows a power-law behavior $\sigma \propto \lambda^\alpha$ with $\alpha < 0$. The inset shows α upon decreasing the broadening η down to about 0.4 meV, which is the present limit of our numerical resolution. Note that the observed behavior is consistent with the limit $\alpha = -2$ for $\eta \to 0$, which has been observed experimentally [15] for the localization of ZEMs by means of scanning tunneling spectroscopy. The localization at $E = 0$ is even stronger than in the Anderson regime and can therefore not be attributed to multiple scattering and quantum interference effects, i.e. the strong localization regime, but is rather a signature of zero-energy modes. This is further corroborated by the length $\lambda \sim 5$ nm over which σ localizes, which is on the same order as the spatial extension of the bound states experimentally measured [15].

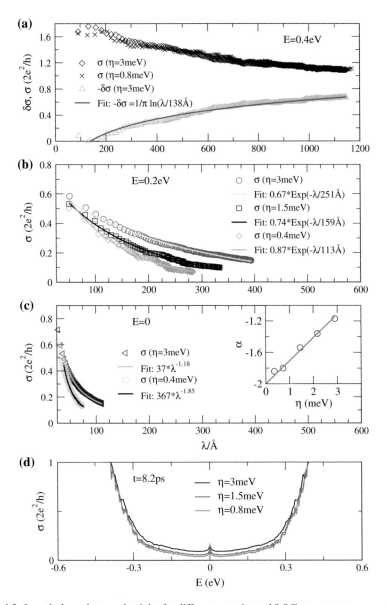

Fig. 4.3 Length-dependent conductivity for different energies and 0.8 % vacancy concentration in the compensated case. **a** Conductivity σ and quantum correction $\delta\sigma = \sigma - \sigma_{sc}$ at $E = 0.4$ eV. The logarithmic fit confirms the weak-localization regime. **b** Low energy conductivity ($E = 0.2$ eV) and corresponding fit indicate Anderson localization regime. **c** At zero energy the conductivity decay is even stronger and cannot be fitted with an exponential decay. **d** Conductivity at largest simulated times (8.2 ps) and its residual dependence on η

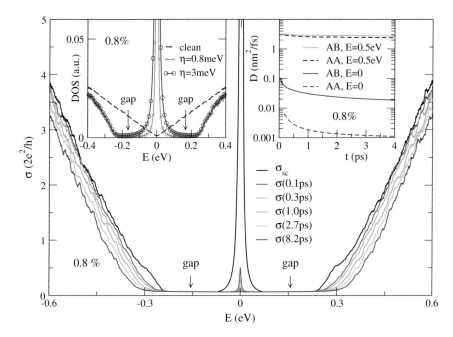

Fig. 4.4 *Main frame* $\sigma_{sc}(E)$ and $\sigma(E, t)$ for graphene (uncompensated case) and energy resolution $\eta = 3$ meV. *Left inset* DOS with energy gap revealed by η scaling and ZEMs. *Right inset* Diffusion coefficients at $E = 0.5$ eV and $E = 0$ ($\eta = 3$ meV) for both compensated (AB) and uncompensated (AA) cases. All data for $n = 0.8\%$

I would like to point out here that our results for compensated vacancies are well-defined and converge in the limit of small η. Figure 4.3d finally shows that at the largest time considered for the calculation of the conductivity (8.2 ps), $\sigma(E)$ is well controlled when decreasing η, with a more pronounced noise level at smaller η, an effect which defines a lower limit for η to avoid non-physical mathematical singularities.

A remarkably different picture emerges in the uncompensated case, for which the sublattice symmetry is fully broken. The DOS shown in Fig. 4.4 (left inset) evidences the presence of ZEMs sharply peaked at $E = 0$. In contrast to the compensated case, the depletion of the low-energy conductivity is here inherited from the presence of energy gaps [16, 17]. The semiclassical conductivity strongly increases when approaching the Dirac point, much more than in the compensated case and also increases when improving the energy resolution. However, the large value of σ_{sc} does not reflect the extendedness of the corresponding ZEMs. This can be rationalized by scrutinizing $\sigma(E = 0, t)$ and $D(E = 0, t)$, which are actually strongly decaying with time. Indeed $D(E = 0, t)$ becomes extremely small compared to that at finite energies (e.g. at 0.5 eV) and much smaller compared to the compensated case with same vacancy concentration (see Fig. 4.4, right inset). Additionally, $D(E = 0, t)$ decays when improving the energy resolution (not shown here), thus demonstrating that although many ZEMs are present, they do not participate in conduction, and that the large value of σ_{sc} obtained numerically results from the high DOS at $E = 0$. Furthermore, the physical relevance of a semiclassical conductivity at the Dirac point is highly questionable. At the Dirac point, we observe that the semiclassical conductivity diverges with small η. The reason is that, for the uncompensated case, all vacancy-induced modes are exactly at $E = 0$ and their corresponding DOS and semiclassical conductivity have a δ-like distribution centered in the gap where no propagation is possible. However, the broadening and the height of the DOS peak (as well as σ_{sc} peak) are artificially driven by the finite parameter η. For the quantum conductivity, the strong decay of $\sigma(E = 0, t)$ with time is consistent with localized modes similar to the compensated case. We also find that away from the Dirac point a higher energy resolution reduces σ_{sc} and $\sigma(t)$ as observed for the DOS, thus unambiguously indicating the energy gap as the origin of the conductivity decrease, and ruling out any diffusive regime and Anderson localization phenomenon. Finally, for larger energies away from the gap region, one observes that the wave packet dynamics for the compensated (AB) and uncompensated (AA) case are very similar, see Fig. 4.4 (right inset). This discards any singular transport mechanism in uncompensated situation, differently to previous reports on hydrogenated graphene [47].

ZEMs effects in disordered finite graphene strips. In contrast to 2D graphene, the role played by ZEMs in transport through finite strips in between highly doped contacts turns out to be quite different. In this configuration, the contacts have much higher density of propagating states than the central strip, especially at the Dirac point. Accordingly, many states from contacts tunnel through the strip as evanescent modes, yielding a minimum ballistic value $\sigma_{min} = 4e^2/\pi h$ for clean samples [1, 25, 26]. The presence of ZEMs increases the number of available states at the Dirac point in the central strip. Two competing transport mechanisms then drive the conductivity behavior, namely an enhanced tunneling probability assisted by ZEMs together with multiple scattering and quantum interferences, which develop owing to the randomness of vacancies distribution.

Figure 4.5 (main frame) shows the quantum conductivity σ for a strip with length $L = 15$ nm, width $W = 150$ nm and compensated vacancy density in the range $[0\%, 2\%]$. In the absence of vacancies, σ shows the minimum conductivity $\sigma(E = 0) \equiv \sigma_0 \approx \sigma_{min}$ expected for the ballistic limit when $L \ll W$ (see horizontal dotted line) [25]. For $n = 0.1\%$, the strip length is close to the mean free path, see Fig. 4.2. Therefore, the transport along the strip remains quasiballistic, a fact further confirmed by the smooth decay of σ all over the spectrum except at the Dirac point, where σ keeps a larger value. For higher densities and away from the Dirac point, the decay of $\sigma(E)$ with n is consistent first with the occurrence of a diffusive regime and then with localization phenomena, as revealed by the strongly fluctuating conductivity. Note that despite the few nanometers short mean free path, even for $n = 2\%$ the conductivity remains significant as a consequence of the large number of conductive channels that penetrate the undoped strip. The conductivity around the Dirac point is further scrutinized in Fig. 4.6 (bottom inset) for strips with $L = 15$ nm, $W = 150$ nm and compensated vacancy densities up to $n = 1\%$. To reduce sample-to-sample fluctuations, all the results were averaged over 20 random disordered configurations.

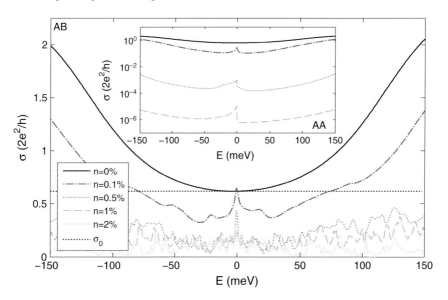

Fig. 4.5 *Main frame* Conductivity for strips with $W = 150$ nm, $L = 15$ nm and compensated vacancy density up to 2 %. *Inset* Same information for uncompensated vacancies with densities up to 1 %

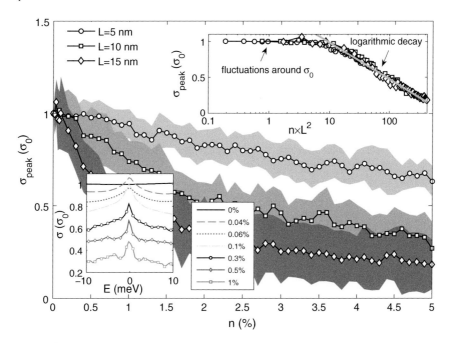

Fig. 4.6 *Main frame* Average conductivity peak versus n for strips with $W = 150$ nm and $L = 5$, 10 and 15 nm. The *shaded areas* around the *curves* indicate the standard deviation with respect to the average value. *Top inset* Same as main frame but as a function of $n \times L^2$. The *thick straight line* is a guide to the eye. *Bottom inset* Average conductivity for $W = 150$ nm, $L = 15$ nm and various n

Far from the Dirac point, the conductivity is found to decrease regularly with n. At $E = 0$, notably enough, a peak is always present, which can slightly exceed σ_0 at very low density ($n \lesssim 0.04\,\%$). This indicates that the ZEMs generated at the Dirac point are sufficiently delocalized to assist (and even enhance) electron tunneling through the strip. Backscattering becomes eventually dominant for sufficiently high defect concentration, as manifested by the smooth conductivity decrease. The dependence of the conductivity peak (σ_{peak}) on the different system parameters is reported in Fig. 4.6 (main frame) for compensated vacancy densities up to 5 % and lengths up to 15 nm. The decrease of σ_{peak} with n is very slow, especially for the shortest strip, and even for strong disorder ($n = 5\,\%$) σ_{peak} remains significantly large. As illustrated in Fig. 4.6 (top inset), σ_{peak} is actually a universal function of $n \times L^2$. Remarkably enough, σ_{peak} fluctuates around or goes slightly above σ_0 for very low $n \times L^2 \lesssim 10$, thus supporting the possibility for a *"supermetallic state"*, introduced by Ostrovsky and coworkers [23, 24]. For $n \times L^2 \gtrsim 10$, σ_{peak} decreases roughly logarithmically, as the result of finite size effects and proximity between vacancies.

The conductivity of graphene strips (with $W = 150$ nm, $L = 15$ nm and n up to 1 %) for uncompensated vacancies are reported in Fig. 4.5 (inset). In marked contrast with the prior case, a gap develops at low density together with reduced but finite conductivity peak at $E = 0$. As for the case of 2D graphene (Fig. 4.4), the gap formation leads to the suppression of tunneling due to the almost vanishing DOS. The Dirac conductivity peak is a signature of the highly localized nature of zero-energy states generated by uncompensated vacancies [17], which are not enough spatially extended to significantly contribute to tunneling and obviate to the DOS decrease.

To further investigate the gap formation as reported in Ref. [17], we consider here the extrinsic density of states, which is given by the difference between the DOS in the presence of vacancies and that for pristine graphene.

Our results for the extrinsic DOS in the compensated (AB) case are plotted in Fig. 4.7a for concentrations from 0.1 to 1 %. We observe that the DOS increases around the Dirac point over an energy region that is larger for higher densities.

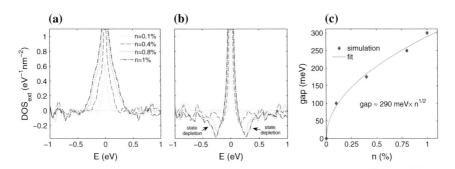

Fig. 4.7 **a** Extrinsic DOS for compensated vacancies as a function of the energy; **b** Same as (**a**) for uncompensated vacancies; **c** Estimation of the gap width and its fit as a function of the density of uncompensated vacancies

Outside this region, the extrinsic DOS fluctuates around 0, meaning that the total DOS is not significantly modified with respect to the clean case. Although the DOS seems to increase considerably in correspondence to the Dirac point, as in [17] our numerical resolution is clearly not good enough to investigate what happens exactly at $E = 0$.

The extrinsic DOS in the uncompensated (AA) case are plotted in Fig. 4.7b, for the same vacancy densities. As expected, the breaking of A-B symmetry generates a relatively sharp peak at zero energy. The peak height increases with vacancy concentration and this occurs at the expense of the DOS at the sides of the Dirac point, where the extrinsic DOS becomes negative. Although we cannot yet be conclusive about this point, it could be the effect of a gap opening, partially hidden by the wings of the convoluted zero-energy peak. This could explain contradictory observations as reported in [17, 47]. Reference [17] pinpoints the opening of an energy gap, whereas [47] suggests the absence of localization in the uncompensated case for energies close to Fermi level. Figure 4.7c shows our estimation of the simulated gap against n and its fit, which gives

$$\text{gap} \approx 290 \times \sqrt{n[\%]} \text{ meV} \tag{4.1}$$

in total agreement with Ref. [17].

In both AB and AA cases, vacancies preserve the hole-particle symmetry (chiral symmetry) and affect the electronic structure around the Dirac point, although in a different manner. In the first case the DOS increases, while for the AA distribution there is a depletion of the DOS around Fermi energy and a finite concentration of zero-energy modes in the middle.

In conclusion, the contribution of ZEMs to quantum transport in disordered graphene has been discussed for various transport geometries and sublattice symmetry-breaking situations. Our findings provide a broad overview of the low-energy transport phenomena in graphene in presence of ZEMs, including the formation of an insulating state at the Dirac point, accessible in absence of electron-hole puddles [12]. The role of electron-electron interaction (here neglected), might also play some important role in capturing the full picture and deserves further investigation [48, 49].

4.2 Charge Transport in Poly-G

4.2.1 Introduction

Graphene-based science and nanotechnology have been attracting considerable interest from the scientific community, in view of the numerous possibilities offered by graphene for not only studying fundamental science in two-dimensional (2D) layered structures [9, 50] but also for improving the performance of flexible materials and for its integration into a variety of electrical and optical applications [51–57].

This interest is driven by graphenes superior mechanical strength and stiffness [58], electronic and thermal conductivity [59, 60], transparency [61], and its potential for straightforward incorporation into current silicon and plastic technologies [62, 63].

For large-area graphene, the CVD growth technique is unquestionably the best candidate for achieving a combination of high structural quality and wafer-scale growth [64–66]. Unfortunately, the transfer of graphene to diverse substrates [67, 68] is still a significant challenge for a plethora of applications, including (bio)chemical sensing [69], flexible and transparent electrodes [64], efficient organic solar cells [70], multifunctional carbon-based composites [63], and spintronic devices [71]. Considerable effort is also needed for fine-tuning of the CVD growth process. In particular, the produced graphene is typically polycrystalline in nature, consisting of a patchwork of grains with various orientations and sizes, joined by GBs of irregular shapes [72, 73]. The boundaries consist of an approximately one-dimensional (1D) distribution of non-hexagonal rings [72, 73], and appear as structural defects acting as a source of intrinsic carrier scattering, which limits the carrier mobility of wafer-scale graphene materials [74].

GGBs also introduce enhanced chemical reactivity [75]. This opens a hitherto unexplored area of research, namely, GGB engineering of the properties of Poly-G, with further diversification of material performance and functionality. Selective chemical functionalization of GGBs with various functional groups and selective adsorption of various metal particles not only modify the carrier mobility of Poly-G but also make it biochemically active, a feature which could be utilized in highly sensitive biochemical sensors. With the capability of engineering GGBs during CVD growth and their applications mentioned above, a new multidisciplinary field of science and engineering can be established. Although graphene oxide is another category of graphene with strong chemical functionalization, the materials exist in a powder form and their use is also different from large area CVD-grown graphene. The extensive review on this has been published elsewhere [76–81]. We limit our discussion to large-area CVD-grown Poly-G here. In this section, we present the current progress of this field through an overview of the experimental efforts to understand the fundamental connection between the structure and the corresponding mechanical, electrical, and chemical properties of Poly-G. I also show why nanotechnology and related methods are essential not only for observing and analyzing GGBs, but also for tailoring nanomaterials with superior performance.

4.2.2 Structure and Morphology of GGBs

4.2.2.1 GGBs Formed Between Two Domains with Different Orientations

While a detailed description of graphene defects has been extensively reviewed already [82–85], here I point out and update some important features of GGB structures. This will aid in understanding the physical and chemical properties of GGBs,

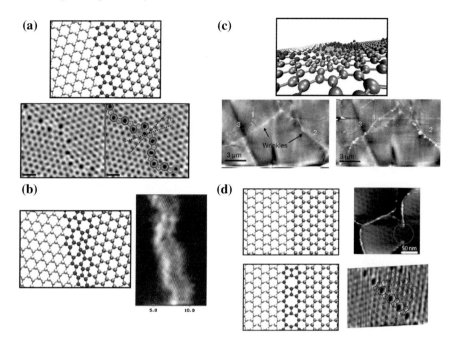

Fig. 4.8 Structure and morphology of GGBs by theory, TEM, and STM/AFM. **a** *Top panel* 5-7 GGB between two graphene grains with a misorientation angle of 21.8°. *Bottom panel* TEM image [72] of a thin 5-7 GGB between grains with a misorientation angle of 27°. Reproduced with permission [72]. Copyright 2011, Nature Publishing Group. **b** *Left panel* simulated construction of a disordered GGB, including a range of non-hexagonal rings and carbon vacancies [86]. *Right panel* STM image of a disordered GGB revealing a similar morphology to the simulated one. Reproduced with permission [87]. Copyright 2012, AIP Publishing. **c** *Top panel* 3D morphology of a 5-7 GGB, indicating out of plane relaxation [88]. *Bottom panels* buckled AFM morphology of Poly-G after UV exposure. Position 2 indicates out of plane buckling at the GGB [89]. Reproduced with permission [89]. Copyright 2012, Nature Publishing Group. **d** The simulated patterns and STM images of two merged grains with identical orientation on a BN substrate (*top panels*) and a Ni substrate (*bottom panels*) [22, 90]. No GGB is observed on the BN substrate, while a 5-8-5 GGB line appears on the Ni substrate. Reproduced with permission [22, 90]. Copyright 2013 and 2010, Nature Publishing Group

with an aim toward controlling their behavior and functionality. GGBs are formed at the stitching region between two graphene domains with different orientations or with a spatial lattice mismatch. In general, a GGB is a thin meandering line that consists of a series of pentagonal, hexagonal, and heptagonal rings [72–74], where the structure and periodicity of the GGB are determined by the misorientation angle between two domains. An example of this is shown in the top panel of Fig. 4.8a, which depicts a 5-7 GGB formed between two grains with a misorientation angle of 21.8°. This GGB consists of a periodic series of pentagon-heptagon pairs. In comparison, the bottom panel of Fig. 4.8a shows a high-resolution transmission electron microscopy (TEM) image of a GGB between two domains with a misorientation angle of 27°. While the experimental image indicates a non-straight GGB, it also consists of a single thin line of pentagon-heptagon pairs [72].

However, this simple GGB structure is not always achieved during the CVD growth process. For example, Fig. 4.8b shows a theoretical model (left panel) and observation by scanning tunneling microscopy (STM; right panel) of a disordered GGB consisting of a complex and meandering series of various carbon rings, as well as the occasional vacancy defect [86, 87]. In this type of structure, the electronic effect of the GGB can extend to several nanometers in width, as can be directly observed from the STM image. Its corresponding transport properties are independent of the orientation of the two domains forming the GGB [86]. In order to minimize the structural energy due to the presence of non-hexagonal rings, the GGB and the surrounding graphene grains can lead to buckling along the length of the GGB [88, 89]. This is true even in the ideal case, and thus is a common feature of all GGBs. For example, the top panel of Fig. 4.8c shows the morphology of a three-dimensional (3D) model of a GGB and its neighboring grains, indicating that out-of-plane buckling can occur. The bottom panel of Fig. 4.8c shows buckled graphene morphology on copper measured before and after ultraviolet (UV) treatment [89]. The buckling line at position 2 coincides with the buckled GGB visualized after UV exposure.

The existence of GGBs can strongly alter the mechanical properties of Poly-G. While monocrystalline graphene has been established as the strongest material ever measured, with an intrinsic strength of $42\,\mathrm{Nm}^{-1}$, a failure strain of 0.25, and a Youngs modulus of 1 TPa [58], the mechanical properties of Poly-G remain under intense scrutiny. The usual method for estimating the elastic properties of 2D materials is to transfer the membrane onto a substrate with an array of holes, and apply a force to the membrane through one of the holes with an atomic force microscope (AFM) [58]. The first reported measurements indicate that GGBs in CVD-grown graphene significantly lower the elastic constant by a factor of six [72, 91, 92], with an average breaking load of about 120 nN, an order of magnitude lower than for monocrystalline graphene [58]. The strength of individual GGBs was also found, theoretically and experimentally, to strongly depend on the misorientation angle between graphene domains [93–97]. However, these results are for a single GGB between two domains, and it is uncertain how they translate to macroscopic samples containing several GGBs. Moreover, the cracks that appear upon failure do not necessarily follow the GGBs but can also penetrate through the grains [89, 98], even if they originate at the GGB regions.

A more realistic model for Poly-G can be constructed by simulating seeded growth of separated graphene grains with random orientations, and allowing such grains to merge together to form natural GGBs [86]. For these samples, the angle-dependence of the mechanical properties vanishes, and clear trends appear as a function of the average grain size. Increasing grain sizes lead to lowering fracture strain and increasing elastic modulus, whereas the variation in the strength of the material is much less affected, being about 50 % of that of monocrystalline graphene [86]. The cracks originate at GGB junctions, and propagate through the grains, in agreement with the experiments [98]. More restricted models containing several connected hexagonal graphene grains have recently confirmed these findings [99].

Although much progress has been made in understanding the mechanical strength of Poly-G, questions still remain. For example, the breaking loads for early measurements [72, 92] differ significantly from those measured more recently [96, 97]. In addition, as noted above, the applicability of the AFM measurements to macroscopic samples remains an open question. To finally resolve the issue, we would need a new measurement technique for estimating the elastic properties of 2D materials, which would avoid the shortcomings of the method utilizing an AFM tip.

4.2.2.2 GGBs Formed Between Two Domains with the Same Orientation

In addition to degraded mechanical properties, numerous studies have shown that carrier transport in Poly-G is strongly affected by GGBs [21, 89, 100–102]. Therefore, a great deal of effort has been made to eliminate the formation of GGBs during CVD by growing monocrystalline graphene [103–108]. There are two primary methods to obtain monocrystalline graphene with CVD. One method is to control the number of nucleation seeds (and thus the individual grain size) by polishing the copper substrate [108], annealing it at high temperature before growth [103, 104], or using copper oxide [105, 106]. Recently, this approach has been able to realize CVD growth of individual grains on the order of several millimeters in diameter. The drawback of this method is that it takes a long time (for instance 12 h) for a single graphene grain to grow to a large size. Furthermore, the crystallinity within a single domain is not guaranteed or at least not confirmed rigorously. Another method is based on controlling the orientation of graphene domains, such that their crystal lattices are aligned [109–112]. One would then expect that these domains will merge cleanly, without forming any GGBs at the stitching regions, as shown in the upper left panel of Fig. 4.8d. However, experiments have shown that this is not always the case. For example, no GGBs were found in the case of graphene growth on a monocrystalline boron nitride (BN) flake [90] (red circle, top right panel of Fig. 4.8d). On the other hand, a line of $5 - 8 - 5$ rings was observed for graphene grown on nickel (Ni; bottom panels of Fig. 4.8d) even though the graphene domains have the same orientation [22]. This is caused by a translational mismatch between neighboring grains. In addition, non-straight edges can also lead to more complex GGB structures than the $5 - 8 - 5$ example shown here. Therefore, additional proof such as high resolution STM, TEM, or electrical transport measurements are necessary to confirm the absence of GGBs in these samples. Different methods of observing GGBs are described below.

4.2.3 Methods of Observing GGBs

To study the properties and structure of GGBs, or to control the graphene growth process, it is necessary to develop methods to determine the location of the GGBs. This information is not straightforward to obtain due to the atomic width of the

GGBs (on the nm scale), and is even more challenging for large-scale observations. A primitive approach is to stop the CVD process before graphene growth is complete. Then, the GGB location can be roughly estimated as the stitching region between two domains [108]. However, graphene domains are not typically monocrystalline and thus a large number of GGBs can be missed with this approach [89, 100, 108].

Another approach to determine the location of the GGBs relies on mapping the orientation of the graphene grains; the shape of each grain is identified, and the GGB locations are then indirectly determined at their boundaries. The techniques for determining the grain orientation include TEM [72, 73], low electron energy microscopy [112], and polarized optical microscopy (POM) of spin-coated liquid crystals on graphene [113–115]. However, these methods will not reveal boundaries between grains with the same orientation. An alternative method, which sidesteps this problem, is to directly observe the location of the GGBs by taking advantage of their chemical properties [83]. These methods are discussed in more detail below.

4.2.3.1 TEM

The principle of using TEM to map the graphene grain orientations is shown in Fig. 4.9 [72]. The diffraction pattern of monocrystalline graphene is six-fold symmetric, corresponding to the symmetry of the honeycomb lattice. If the observed region includes two different orientations, the diffraction pattern consists of two different hexagons rotated by a specific angle, as shown in Fig. 4.9a. This is the misorientation angle between the two grains. By doing this analysis over the entire

(a)   **(b)**

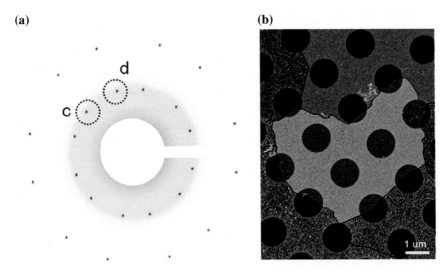

Fig. 4.9 TEM approach to identifying graphene grain orientations. **a** An electron diffraction pattern arising from two misoriented grains. **b** Mapping of several grains with different orientations. Reproduced with permission [73]. Copyright 2011, ACS Publishing

sample, one can map the orientation of the graphene lattice at each point in the sample. An example is shown in Fig. 4.9b, where the colored regions mark grains of different orientations.

4.2.3.2 Liquid Crystal Deposition

Although TEM observations provide atomic resolution of GGBs at a nanometer scale, the GGB distribution at millimeter or centimeter scales is not easily accessible. Here, we describe several methods of observing GGBs at large scale. Figure 4.10a shows the principle of using liquid crystal (LC) (4-Cyano-4pentylbiphenyl; 5CB) molecules to observe graphene grain orientation with POM [114]. A 5CB molecule consists of two hexagonal benzene rings with a nitrogen atom at one end and a long carbon chain at the other end. It is expected that the hexagonal rings of the 5CB molecule will align along the graphene lattice with AB stacking order. Graphene grains with different orientations provoke the 5CB molecules to align in different directions depending on the grain orientation, which can be observed as a contrast difference using POM. This can be seen in Fig. 4.10b, which shows two POM micrographs that indicate a

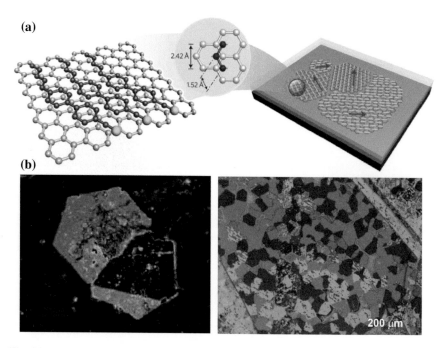

Fig. 4.10 Liquid crystal coating approach to identifying graphene grain orientations. **a** The hexagonal rings of LC molecules align coherently with hexagonal rings in graphene. Reproduced with permission [114], Copyright 2012, Nature Publishing Group. **b** POM images of LC molecules aligned on each graphene grain, revealing a strong optical contrast between misoriented grains

clear contrast between graphene grains of different orientations. This approach can be extended to a large scale, as shown in the right panel of Fig. 4.10b. Interestingly, experiments have not revealed a three-fold symmetry for the alignment of the 5CB molecules on graphene, which would be theoretically expected. Further studies are required to fully understand the rearrangement of LC molecules.

4.2.3.3 UV Treatment

Instead of mapping the orientation of each graphene grain, the high chemical reactivity of GGBs can be utilized for their direct visualization [89, 116, 117]. One approach involves the use of an oxidizing agent to selectively oxidize copper underneath the GGBs [89]. Figure 4.11a shows the principle of UV treatment of graphene on a copper substrate in a humid environment. O and OH radicals are generated under UV exposure, and these radicals can easily invoke strong chemical reactions near the defect sites. In particular, GGBs, aggregates of defects such as vacancies, pentagons and heptagons, are most vulnerable for radical attack. These radicals penetrate through graphene defects at the GGBs to oxidize the underlying copper substrate,

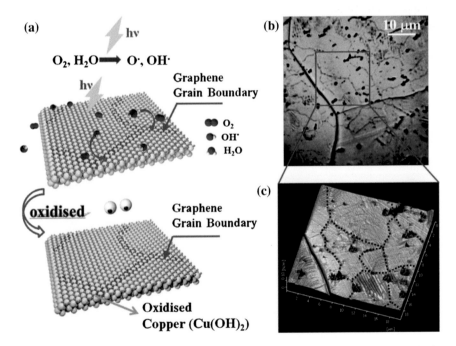

Fig. 4.11 UV treatment approach to identifying graphene grain orientations. **a** Principle of GGB visualization by UV treatment. **b, c** Selective oxidation of an underlying the copper substrate for direct optical identification (**b**) of the GGBs, confirmed by AFM (**c**). Reproduced with permission [89]. Copyright 2012, Nature Publishing Group

forming copper oxides. This provokes volume expansion to several hundred nm in the region of the GGB lines, and these oxidized lines can then be observed under an optical microscope. Figure 4.11b, c are optical and AFM images of the graphene sample after UV treatment, clearly indicating the positions of the GGBs.

It is worth noting that the methods discussed in this section are complementary to each other, where a combination of techniques can be used to visualize GGBs from the atomic scale to the wafer scale. LC coating and overlapping two graphene layers can easily determine the location of GGBs when the grains have different orientations. However, it is not possible to use these methods to determine if two grains have the same orientation. In this case, TEM, STM, or the UV oxidation methods are required.

4.2.4 Transport Properties of Intrinsic Poly-G by Simulation

In addition to their structural characterization and identification, it is important to understand how the GGBs influence electrical transport phenomena in Poly-G. Here, we provide a comprehensive theoretical picture of the relationship between a poly-crystalline morphology and the resulting charge transport properties. We explored large models (up to 278,000 atoms) of intrinsic Poly-G samples with varying misorientation angles, realistic carbon ring size statistics and non-restricted GB structures. For this purpose, we used an efficient computational approach that is particularly well suited for large samples of low-dimensional systems [32]. We calculated charge mobilities in these samples using a TB Hamiltonian and an efficient real space (order-N) quantum transport method, which enabled us to establish the scaling law for transport properties for samples with well interconnected grains. This scaling property is inferred from the observed electron-hole density fluctuations that develop at the atomic scale along the boundaries. For poorly connected samples, we observed greatly reduced mobilities, which agrees with experimental results [100]. These findings offer unprecedented insight into the transport fingerprints of intrinsic Poly-G samples.

4.2.4.1 Models

Our model structures were created using the method outlined in [118]: (1) Nucleation sites for a selected number of randomly oriented graphene grains are randomly placed on a pre-defined two-dimensional simulation cell; (2) Atoms are randomly added to the reactive sites at the edges of the grains until two grains meet, at which point the growth is locally terminated; (3) When no reactive sites are free, the structure is heated to 3,000 K for 50 ps within a molecular dynamics simulation to allow the GB structures to overcome the most spurious atomic configurations; (4) The structure is quenched during a 10 ps simulation run to enable the lattice to obtain its equilibrium size (zero pressure). Since a prerequisite to the efficient calculation of electronic properties in this study was that the structures had to be flat, at this point, we removed small corrugations which appeared after the aforementioned preparation steps. To this end, the structures were repeatedly stretched, gradually forced towards zero in the third dimension (by scaling down the z-coordinates), and again relaxed (allowing atomic reconstruction at each step), which removed the largest portion of non-flat configurations. A few remaining non-flat and physically implausible configurations (overlapping atoms, coordination numbers higher than three) were removed manually, and a final relaxation and optimization step was carried out. This resulted in flat structures occupying local energy minima and suitable for the present study.

During the sample preparation, the carbon-carbon interactions were modeled using the reactive bond order potential by Brenner et al. [119] and the temperature and pressure control were handled using the Berendsen method [120]. Most of the structures were approximately $60 \times 60\,\text{nm}^2$ in size and contained $\sim$138,000 atoms with the exception of one structure which was significantly larger ($87 \times 87\,\text{nm}^2$, $\sim$278,000 atoms). Structure with the smallest grains contained 22 of them, whereas whereas all other structures contained 11 grains. Periodic boundaries were used in all calculations.

4.2.4.2 The Scaling Law

For electronic and transport calculations, we used a π-π^* orthogonal TB model, described by a single p_z-orbital per carbon site, with nearest neighbors hopping γ_0 and zero onsite energies. A distance criterion to search for the first nearest neighbors was set empirically to $1.15 \times a_{CC}$, where a_{CC} is the nearest neighbor distance in pristine graphene. The local fluctuations in bond lengths are small enough to keep a constant value of γ_0 for the transfer integral. The density of states (DOS) was computed using the Lanczos recursion method with $N = 1{,}000$ recursion steps and an energy resolution $\eta = 0.01\gamma_0 \simeq 0.03$ eV. For LDOS calculations we used the spectral measure operator $\delta(E - \hat{\mathcal{H}})$ projected on state $|i\rangle$ (where i is the site index).

We computed the local charge density deficiency δ_i (or self-doping) for each GB site i defined as:

$$\delta_i = \int_{-\infty}^{E_{itCNP}} [\rho_{tot}(E) - \rho_i(E)]dE \qquad (4.2)$$

where ρ_{tot} and ρ_i are the total DOS of the Poly-G sample and the LDOS on carbon site i, respectively. E_{CNP} denotes the charge neutrality point.

To capture the different transport regimes, we employed a real-space order-N quantum wavepacket evolution approach in Chap. 3 to compute the Kubo-Greenwood conductivity [32]. As has been shown before [118], our models for Poly-G resemble experimentally observed structures: atomic-resolution and diffraction-filtered electron microscopy experiments have revealed that the grains stitch together predominantly via pentagon-heptagon pairs [72, 73, 121] in arrangements of large number of small grains forming an intricate patchwork interconnected by tilt boundaries [72, 73]. For this study, we created samples with three different average grain sizes

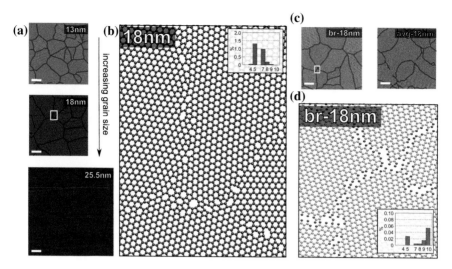

Fig. 4.12 **a** Three structures with uniform grain size distribution and increasing average grain sizes (13.0, 18.0 and 25.5 nm). GBs are marked with *dark lines*. **b** Larger magnification of the *area marked* with a *white rectangle* in panel (**a**), showing a typical example of the GBs. Carbon ring-size statistics for the same sample (showing the ratio of non-hexagonal rings) are presented in the *upper right* corner. **c** Two additional samples with average grain size of 18 nm: one sample with broken boundaries ("br-18 nm") and another one with random grain size distribution ("avg-18 nm"). **d** Higher magnification of the *area marked* with a *white rectangle* in panel (**c**), showing the structure of "broken" boundaries in sample "br-18 nm". The statistics of non-hexagonal rings are shown in the *lower right* corner. All scale bars are 10 nm

(average diameter $\langle d \rangle \approx 13$, 18 and 25.5 nm) and uniform grain size distributions (Fig. 4.12a). As seen in Fig. 4.12b, the atomic structure at the GBs consists predominantly of five- and seven-membered carbon rings and assumes meandering shapes similar to the experimentally observed ones. We also created one sample with $\langle d \rangle \approx 18$ nm and "broken" (poorly connected) boundaries ("br-18 nm"), and one sample with $\langle d \rangle \approx 18$ nm and non-uniform d-distribution ("avg-18 nm") (see Fig. 4.12c, d).

We begin by discussing the electronic density of states (DOS) as a function of energy (E) for the different samples (Fig. 4.13a). We noticed very little variation away from charge neutrality point ($E = 0$), except for a slight broadening of van Hove singularities at $E = \pm \gamma_0$, where $\gamma_0 = -2.9$ eV is the nearest neighbor hopping energy. This suggests that GBs induce weak disorder and that the polycrystalline samples mostly preserve the electron-hole symmetry. However, a larger difference can be seen at the charge neutrality point (Fig. 4.13b), where all of the polycrystalline structures show an enhanced density of zero energy modes [16]. As expected, the largest difference relative to pristine graphene was with the "br-18 nm" sample (the one with poorly connected grains), reflecting a higher density of "midgap" states [16, 17].

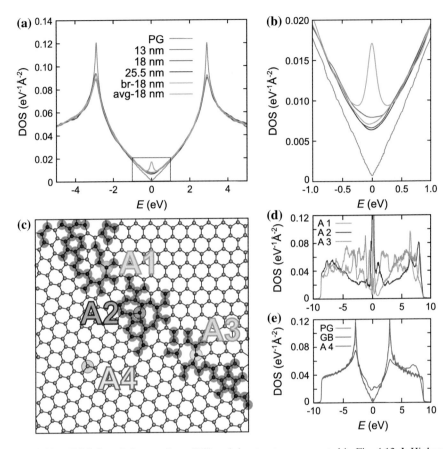

Fig. 4.13 **a** DOS for pristine graphene (PG) and the structures presented in Fig. 4.12. **b** Higher magnification of the DOS close to the charge neutrality point ($E = 0$, *area marked* with a *rectangle* in panel (**a**)). **c** Atomic structure of one of the boundaries in sample "18 nm", showing the electron-hole density fluctuations at GB sites that develop due to local variations in the charge density δ_i: local electron doping ($\delta_i < -1 \times 10^{-4}$ e/atom) is shown in *blue* and local hole doping ($\delta_i > 1 \times 10^{-4}$ e/atom) in *red*. **d** Local DOS for atoms A1, A2 and A3 marked in panel (**c**). **e** Local DOS for atom A4 marked in panel (**c**) as compared to the average DOS for pristine graphene (PG) and average LDOS for all atoms at GBs in the same sample (GB)

To better understand the deviations from the pristine graphene for the well-connected structures, we next identified atoms residing at GBs of the "18 nm" sample by searching for atoms for which the bond length of at least one nearest neighbor differs from the carbon spacing in pristine graphene ($a_{CC} = 1.42$ Å) by 0.03 Å or more. We then calculated the local charge density deficiency δ_i (or self-doping) for each GB site. In Fig. 4.13c we present the atomic structure of the electron-hole density fluctuations (δ_i variations greater than 10^{-4} electrons per atom) formed at a small area around one GB. These self-doping effects stem from local fluctuations in the electrostatic potential. Experiments on exfoliated graphene deposited over silicon dioxide

[122, 123] have shown similar potential inhomogeneities; however, these were spread over a much longer scale (~30 nm) and were induced by proximity effects generated by charges trapped in the oxide. In our case, averaging over all carbon atoms belonging to the GBs of the 18 nm sample gave $\langle \delta \rangle_{GB} = 0.008$ electrons per atom, which corresponds to a mean carrier density of $\langle n(E = 0) \rangle \simeq 6.1 \times 10^{11} cm^{-2}$. ($\delta$ fluctuates between -0.096 and 0.08 electrons per carbon atom, or, respectively, 6.1×10^{12} and -7.3×10^{12} cm^{-2}.) The local charge density fluctuations occur on a length scale only a few times larger than the lattice spacing, which is very small compared to that in supported exfoliated graphene, suggesting a much stronger local scattering efficiency. We point out that our results show no straightforward correlation between the self-doping value and the local defected morphology of the lattice.

Figure 4.13d shows the plot of the corresponding local DOS (LDOS) of three selected atoms at the boundary (A1, A2 and A3). All of them show increased contributions of midgap states [16, 17], significantly reduced van Hove singularities, and a markedly enhanced electron-hole asymmetry, owing to the odd-membered carbon rings [32]. They also exhibit strong resonant peaks, which are characteristic of quasi-localized electronic states in the vicinity of defects. The local electronic configuration along the GB also strongly differs from one site to another, an effect arising from an interference effect between coherent wave functions of the connected adjacent grains. In clear contrast, an atom only four lattice vectors away from the boundary (A4) shows a LDOS nearly indistinguishable from that of the pristine graphene (Fig. 4.13e). Comparison to the average LDOS calculated for *all* atoms at the GBs reveals that the changes in the DOS seen in the polycrystalline samples (Fig. 4.13a) arise locally from the atomic configurations of the GBs itself.

Next, we discuss the transport properties of the samples. Figure 4.14a shows the time dependency of the diffusion coefficient $D(t)$ at the Dirac point for all samples. On the one hand, the well-connected samples display a very slow time-dependent decay of $D(t)$ after the saturation value, indicating weak contribution of quantum interferences. On the other hand, the poorly connected sample "br-18 nm" exhibits a much faster decay, eventually driving the electronic system to a strong localization regime (as observed in some transport measurements [102]). We next deduced the mean free path $\ell_e(E)$ from the maximum values of $D(t)$ (Fig. 4.14b). Genuine electron-hole asymmetry is apparent in $\ell_e(E)$, but only for energies $|E| > 3$ eV (far from the experimentally relevant energy window). At lower energies around the charge neutrality point ($|E| < 1$ eV), $\ell_e(E)$ changes, albeit only weakly, for all samples.

The sample with broken boundaries, "br-18 nm", shows the shortest $\ell_e < 5$ nm and the weakest dependence on energy, except for a pronounced dip at $E = 0$. Interestingly, the curves for the two well-connected samples with similar $\langle d \rangle$ but different d-distributions ("18 nm" and "avg-18 nm") are very similar and clearly different from samples with either smaller or larger grains. However, this difference can by accounted for by a constant factor. Remarkably, it turns out that $\sqrt{2} \times \ell_e^{13 \text{ nm}} \approx \ell_e^{18 \text{ nm}}$ and $\sqrt{2} \times \ell_e^{18 \text{ nm}} \approx \ell_e^{25.5 \text{ nm}}$ (see the scaled values in Fig. 4.14b), which correspond exactly to the differences in the average grain sizes in these samples

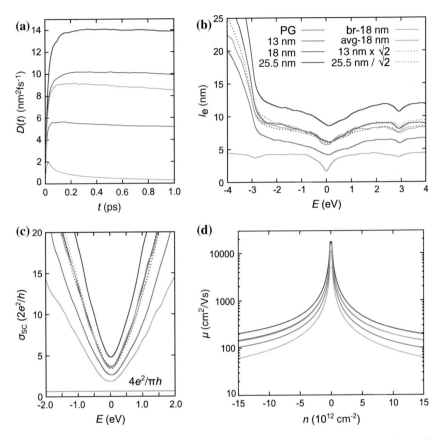

Fig. 4.14 a Diffusion coefficient ($D(t)$) for the samples presented in Fig. 4.12. **b** Mean free path $\ell_e(E)$ for equivalent structures with scaled $\ell_e(E)$ for samples with $\langle d \rangle \approx 13$ nm and $\langle d \rangle \approx 25.5$ nm, showing the scaling law. **c** Semi-classical conductivity ($\sigma_{sc}(E)$) for all samples and as scaled for the same cases as above. **d** Charge mobility ($\mu(E) = \sigma_{sc}(E)/en(E)$) as a function of the carrier density $n(E)$ in each of the samples ($n(E) = 1/S \int_0^E \rho(E)dE$, S being a normalization factor)

($\sqrt{2} \times 13 \approx 18$ and $\sqrt{2} \times 18 \approx 25.5$). Moreover, the grain-size distribution does not enter into this scaling behaviour ($\ell_e^{18\ nm} \approx \ell_e^{avg-18\ nm}$). Hence, we have identified a remarkably simple scaling law that links the average grain size to transport length scales in Poly-G with randomly oriented grains.

The computed semi-classical conductivity $\sigma_{sc}(E)$ exhibits energy-dependent variations similar to $\ell_e(E)$, as can be seen in Fig. 4.14c. We also point out the linear dependency of ℓ_e with charge density in the Dirac point vicinity. Again, the same scaling law (presented above for the mean free path) applies: the ratio of σ_{sc} for two samples with different average grain sizes matches closely with the ratio of the $\langle d \rangle$ values themselves. One additional interesting feature seen in Fig. 4.14c is that the conductivity remains much higher than the minimum value $4e^2/\pi h$ (horizontal line), which fixes the theoretical limit in the diffusive regime, as derived within the

Table 4.1 Mobilities for all samples at selected charge densities

Mobilities (cm^2/Vs)	13 nm	18 nm	Avg-18 nm	25.5 nm	br-18 nm
$\mu(n = 2.5 \times 10^{11} cm^{-2})$	5.1×10^3	7×10^3	6.8×10^3	10^4	4×10^3
$\mu(n = 2.5 \times 10^{12} cm^{-2})$	510	700	685	950	360
$\mu(n = 2.5 \times 10^{13} cm^{-2})$	69	105	104	150	45

self-consistent Born approximation valid for any type of disorder [43]. This indicates that Poly-G remains a good conductor, even for the poorly connected structure "br-18 nm".

Localization length of electron states ($\xi(E)$) can now be estimated using the values for ℓ_e and σ_{sc}. Scaling analysis ($\xi(E) = \ell_e(E) \exp(\pi h \sigma_{sc}(E)/2e^2)$ [13]) reveals that $\xi \simeq 1 - 10\,\mu m$ over a large energy window around the charge neutrality point. This contrasts with the values (on the order of 10 nm) obtained for graphene structures with ~1 % structural defects, strongly bonded adatoms, or other types of short range impurities [32, 33].

Finally, we move on to the charge carrier mobility $\mu(n)$ (Fig. 4.14d). As expected, the poorly connected sample "br-18 nm" shows the lowest mobility (reduced by a factor of about three when compared to the well-connected samples with similar $\langle d \rangle$). We point out that the computed values are valid down to the charge neutrality point (that is, to the smallest charge density $n(E)$), since we accounted for the disorder-induced finite DOS, which yields a non-zero charge density (and thus no singularity at $1/n(E)$). Table 4.1 gives the mobilities at several charge densities for all studied samples. It is worth observing that the scaling law also roughly applies to charge mobilities versus average grain size, since the superimposed effect of density of states changes the ratio only by a few percent (for instance, at $n = 2.5 \times 10^{12} cm^{-2}$, $\mu^{18\,nm}/\mu^{13\,nm} \approx 1.37$).

If we extrapolate the mobility for well connected grains according to our scaling law to a grain size of $1\,\mu m$ and a charge density of $n = 3 \times 10^{11} cm^{-2}$ as in the best samples of Ref. [100], we obtain 300,000 $cm^2 V^{-1}$, which is about ten times higher than the measured values. This discrepancy suggests that substrate-related disorder effects, as well as supplementary defects introduced during the transfer process, should account for an even greater limitation for charge mobilities than the actual GB morphology.

The existence of more disordered GBs as reported in Refs. [87, 100], or samples with overlapping grains, as observed in Ref. [124], yield to lower mobility values, which has been partly illustrated here with the structural model "br-18 nm". More work is however needed to design proper atomistic structural models that will capture essential geometrical features of those more fragmented structures of Poly-G.

In conclusion, we have created Poly-G samples with non-restricted GB structures and realistic misorientation angles and ring statistics. These samples enabled us to confirm the simple relationship between the average grain size and charge transport properties of intrinsic Poly-G. This scaling law will be explained more below in Sect. 4.2.5.3. The disorder scattering strength in Poly-G was found to depend on the atomic structure of GBs (inducing quasi-bound states at resonant energy) and wavefunction mismatch between the grains, which generate strongly fluctuating, but highly localized electron-hole density fluctuations along the interfaces between grains. Our results significantly improve the present theoretical understanding on the influence of the detailed morphology of polycrystalline materials to their measurable electronic properties. They offer the possibility for estimating charge mobilities in suspended CVD-graphene samples based on the average grain sizes and quality of the GBs. Furthermore, they establish quantitative foundations for estimating the intrinsic limits of charge transport in Poly-G, which is of prime importance for graphene-based applications in the future.

4.2.5 Measurement of Electrical Transport Across GGBs

Various measurements have been made to understand the electrical properties of GGBs. These measurements fall into three primary approaches. The first approach involves local two-point measurements, which are accomplished with STM and scanning tunneling spectroscopy (STS) [87, 125–129]. With these measurements, it is possible to deduce the local electronic density of states, the local charge density, and the charge scattering mechanisms associated with GGBs, thus permitting the spatially-dependent electrical characterization of GGBs at the atomic scale. The second approach involves four-probe measurements, which can be used to analyze the influence of individual GGBs at a scale of several micrometers [100–102]. By subtracting the contribution of each graphene grain from an inter-grain resistance measurement, the resistivity of a single GGB can be estimated. In combination with microscopic or spectroscopic techniques, this approach allows one to correlate the resistivity of a single GGB with its structural or chemical properties. Finally, the global impact of GGBs can be studied by measuring the sheet resistance of Poly-G samples over a wide range of average grain sizes and distributions, which are tunable by the CVD growth conditions. By employing a simple scaling law (as discussed below), it is then possible to extrapolate the average GB resistivity [89, 130]. Taken together, these measurement techniques provide the electrical characterization of GGBs at various length scales, thus helping to reveal a comprehensive picture of charge transport in Poly-G. A more detailed overview of these methods is given below.

4.2.5.1 Two-Probe Measurements

Two-probe STM and STS techniques can be used to locally study the electrical properties of GGBs [87, 125, 127–129]. By varying the voltage and position of the STM tip, it is possible to determine the nature of localized states, the charge doping, and the local scattering mechanism corresponding to a given morphology of the

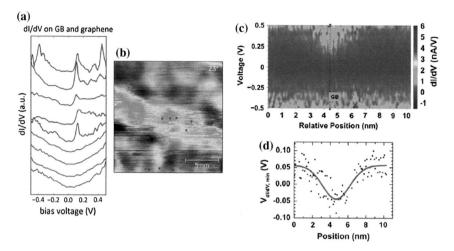

Fig. 4.15 Two-probe measurement of GGBs. **a** Differential tunneling conductance at various points on (*blue lines*) and around (*red lines*) a GGB. The appearance of defect states is evident on the GGBs. Reproduced with permission [129]. Copyright 2013, Elsevier Publishing. **b** STM image of the GGB studied in panel (**a**) where the *colored dots* indicate the positions of dI/dV measurements. **c** dI/dV map across a GGB. **d** Location of the dI/dV minimum as a function of tip position, indicating the presence of an electrostatic barrier at the GGB. Reproduced with permission [128]. Copyright 2013, ACS Publishing

GGB. One example of such analysis is shown in Fig. 4.15a, b [129]. Figure 4.15a shows the differential tunneling conductance, dI/dV, taken at various points on (blue curves) and next to (red curves) a GGB in CVD-grown graphene. A STM profile of the GGB and the points where the measurements were made is shown in Fig. 4.15b. These results indicate the presence of a peak in the tunneling conductance near the Dirac point whenever the STM tip lies on top of the GGB. Meanwhile, this peak does not appear for measurements away from the GGBs. Density functional theory (DFT) calculations have attributed this peak to the localized states arising from two-coordinated carbon atoms in the GGBs [129]. The STM map (not shown here) also reveals interference superstructures due to scattering from the GGBs, indicating the contribution of significant inter-valley scattering. This supports the hypothesis about the presence of two-coordinated atoms, since inter-valley scattering stems from atomic-scale lattice defects [35].

Figure 4.15c shows another map of dI/dV curves as the STM tip is scanned across a GGB [22]. Similar to Fig. 4.15a, an enhanced local density of states is observed at positive voltage when the tip is located over the GGB. The voltage associated with the minimum of dI/dV, as shown in Fig. 4.15d, indicates a strong negative shift around the position of the GGB, revealing n-type doping of the GGB compared to bulk p-type doping of the graphene grains. This shift in doping corresponds to an electrostatic potential barrier of a few tens of meV. Finally, STM interference patterns indicate that some GGBs are dominated by inter-valley scattering while others are dominated by backscattering. The type of scattering appears to depend on the structure of the GGB, where a GGB consisting of a continuous line of defects shows primarily backscattering behavior and a periodic line of isolated defects is dominated by inter-valley scattering.

Other STM studies of GGBs reveal similar results to those mentioned above, with GGBs forming $p - n - p$ or $p - p' - p$ junctions with the bulk-like graphene grains, where $p' < p$. The doped regions associated with the GGBs are on the order of a few nm wide, showing an abrupt transition between the GGB and the grain [87, 128]. Other works reveal the presence of localized states along GBs in graphene and graphite [22, 125, 131]. In general, STM/STS studies indicate that GGBs are a source of localized states and electrostatic potential barriers in Poly-G, and can serve as significant sources of charge scattering.

4.2.5.2 Four-Probe Measurements

In order to make a four-probe measurement of the resistivity of a GGB, it is necessary to first identify its location. This can be done, e.g., with non-destructive TEM or by drop-casting a liquid crystal layer [100, 114, 115]. In the case of two regular hexagonal graphene domains merged together, simple optical microscopy can also be used to identify the boundary location, as shown in the grey background of Fig. 4.16. A Hall bar is then fabricated by e-beam lithography, and a regular four-probe measurement is performed to determine the resistance of the left (L) domain, the right (R) domain,

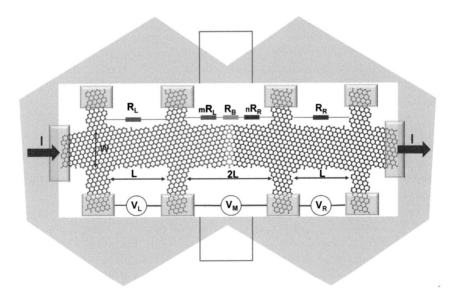

Fig. 4.16 Principle of four-probe measurement applied to GGBs. A serie of Hall bars is fabricated across the GGB region. The resistivity of the GGBs can be extracted from this measurement set-up Intra-grain resistances RL and RR are subtracted from the inter-grain resistance to obtain RB, the resistance of the GGB

and the middle (M) region between the two domains. A constant current is applied from the left to the right while the voltage drop between two adjacent electrodes is measured, and the resistance is calculated by Ohms law, $R_L = V_L/I$, $R_R = V_R/I$, and $R_M = V_M/I$. In general,

$$R_M = mR_L + R_B + nR_R = \alpha R_D + R_B \qquad (4.3)$$

where $m + n = \alpha$ (due to the αL length of the middle part) and R_D is average resistance of the graphene domains, $R_D = \frac{mR_L + nR_R}{m+n}$. If the samples are uniform ($R_L = R_R = R_D$) or if the GGB is located precisely in the middle ($m = n$), then the resistance of the GGB is determined. Otherwise, the precise location of the GGB needs to be determined to extract its resistance. The resistivity of the GGB (ρ_{GB}) is calculated from [100]

$$R_M = \alpha R_D + \frac{\rho_{GB}}{W} \qquad (4.4)$$

Note that ρ_{GB} has the same dimensions as bulk resistivity (Ωm). The relationship between ρ_{GB} and bulk resistivity ρ_{GB}^{bulk} is

$$R_B = \frac{\rho_{GB}}{W} = \frac{\rho_{GB}^{bulk} . l_{GB}}{t.W}, \qquad (4.5)$$

where l_{GB} and t are the effective width and thickness, respectively, of the GGB.

As described above, four-probe measurements are a useful tool for addressing the electrical transport properties of individual GGBs. With this measurement technique, the contribution of within the grains can be separated from the inter-grain resistance, and by normalizing for the length of the GGB, the characteristic transverse GGB resistivity ρ_{GB} is derived. These measurements also yield useful information about the performance of devices based on CVD graphene, because the measurements are made in a device configuration. An example of the experimental setup and measurement results can be seen in Fig. 4.17a, b [101, 102]. Figure 4.17a is an optical image of the four-probe measurement setup across an approximately 4-μm-long GGB. Figure 4.17b shows the I-V curves corresponding to the left and right grains (red and blue curves) and across the GGB (green curve). Here, the I-V curves indicate a much larger inter-grain resistance compared to the resistance measured within each grain, indicating extra scattering provided by the presence of the GGB. This particular measurement yielded a GB resistance of 2.1 kΩ, or $\rho_{GB} = 8$ kΩ.μm when scaled by the GGB length. Temperature-dependent measurements show that ρ_{GB} is insensitive to temperature, pointing to a defect-induced scattering mechanism. Magnetotransport measurements reveal the presence of WL at low temperatures [101, 102], indicating that GGBs are significant sources of inter-valley scattering, in agreement with the STM studies mentioned above.

A similar measurement setup is shown in Fig. 4.17c, on a device fabricated on a specially prepared TEM window that allows for concurrent transport measurements and identification of the individual grains and the GGB [100]. An example of the

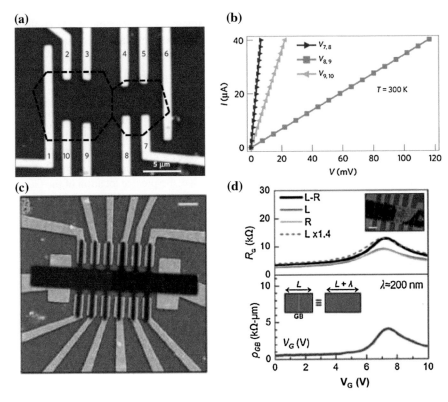

Fig. 4.17 Four-probe measurement of GGBs. **a** Example of a four-probe setup for measuring the resistivity of a GGB. **b** I-V curves measured within individual grains (*red and blue curves*) and across the GGB (*green curve*). The reduced slope for the inter-grain measurement indicates extra resistance contributed by the GGB. Reproduced with permission [102]. Copyright 2011, Nature Publishing Group. **c** Four-probe measurement setup mounted on a TEM holder, where individual graphene grains are identified in the *red and blue* regions. **d** *Top plot* four-probe measurements of the inter- and intra-grain resistance as a function of gate voltage (*black and gray curves*, respectively). *Bottom plot* the extracted GB resistivity as a function of gate voltage in volt. Reproduced with permission [100]. Copyright 2013, AAAS (color in online)

measurement results can be seen in Fig. 4.17d. In the top graph, the gray curves correspond to the resistance measured within each grain, while the black curve is the inter-grain resistance. In the bottom graph, the green curve shows the extracted GB resistivity as a function of applied gate voltage. Here, ρ_{GB} peaks at a value of 4 $k\Omega.\mu m$ at the Dirac point. With the four-probe measurements, ρ_{GB} has been extracted for CVD graphene prepared under several growth conditions, and it has been shown that the resistivity depends strongly on the structure of the GGB. For example, a growth procedure yielding well-connected grains gives $\rho_{GB} = 1$ to 4 $k\Omega.\mu m$ at the Dirac point, while a growth procedure yielding poorly stitched grains results in values of ρ_{GB} an order of magnitude larger. Interestingly, some overlapping GBs have a negative resistivity, with the inter-grain resistance smaller than the combined resistance of the individual grains. This is attributed to reduced scattering in the double-layer overlapped region compared to the single-layer grains.

4.2.5.3 Global Measurements from Scaling Law

In general, GGBs are formed randomly during the CVD growth process, and their electrical properties are not uniform. Therefore, in addition to studies of individual GGBs, it is also necessary to study GGBs on a large scale to extract a reliable average of their transport properties. This average quantity is represented by the GB resistivity ρ_{GB}, which can be extracted from an Ohmic scaling law, as illustrated in Fig. 4.18. Figure 4.18a shows a 1D model of n graphene grains separated by n GBs. The sample resistance R includes the resistance of the n grains R_i^G, and the resistance of the n GGBs R_i^{GB} ($R = \sum_{i=1}^{n} R_i^G + \sum_{i=1}^{n} R_i^{GB}$). These terms can be written as $R = R_S.L/W$, $R_i^G = R_{S,i}^G.L_i/W$ and $R_i^{GB} = \rho_i^{GB}/W$, where R_S is the overall sample sheet resistance, $R_{S,i}^G$ is the sheet resistance of each grain, ρ_i^{GB} is the resistivity of each GB, and L_i is the length of each grain. Putting all this together, the sample sheet resistance can be written as $R_S = \sum_{i=1}^{n} R_{S,i}^G \cdot \frac{L_i}{L} + \sum_{i=1}^{n} \frac{\rho_i^{GB}}{L}$. The first term is the average sheet resistance of the graphene grains R_S^G, which is independent of n, while the last term strongly relies on n or the grain size. This term is equivalent to $n\rho_{GB}/L = \rho_{GB}/l_G$, where ρ_{GB} is the average GB resistivity and l_G is the average grain diameter. The final expression is

$$R_S = R_S^G + \rho_{GB}/l_G \tag{4.6}$$

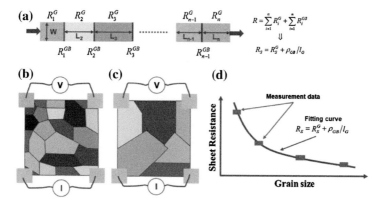

Fig. 4.18 Principle of the scaling law to extract the GGB resistivity. **a** Derivation of the ohmic scaling law. **b, c** Sheet resistance measurements of graphene with small and large grain sizes. **d** Extraction of GGB resistivity by fitting the scaling law to sheet resistance measurements

which is the scaling law in Sect. 4.2.4.2 for the case of completely clean graphene grains ($R_S^G = 0$). R_S can be measured by the Van der Pauw method, as shown in Fig. 4.18b, c. The average grain size can be estimated by visualizing the GB structure of the sample or with Raman measurements, as described in the main text. By measuring the sheet resistance of samples that span a range of average grain sizes, one can extract R_S^G and ρ_{GB}, as shown in Fig. 4.18d.

The two- and four-probe measurement techniques yield valuable information about the electrical properties of GGBs at the atomic and individual-grain scales. These microscopic electrical properties can be correlated to the macroscopic ones, which are applicable to the analysis of experimentally available large-area graphene. This can be accomplished with the global scaling law, as discussed above. Two examples of this procedure are given in Fig. 4.19a, b. Figure 4.19a shows a series of sheet resistance measurements over several orders of magnitude of average grain size [35, 130, 132–135]. Applying the scaling law to this data (black line in Fig. 4.19a results in $\rho_{GB} = 0.67$ kΩ.μm. This value is somewhat lower than those obtained in the four-probe measurements mentioned above. However, because the measurements did not involve back gate modulation, it is likely that the sheet resistance was measured away from the Dirac point, resulting in a lower value of ρ_{GB}. It should also be noted that the x-axis of Fig. 4.19a was obtained through the D/G ratio in Raman spectroscopy, and thus represents an average distance between defects rather than the true grain size.

Another example of the scaling behavior is shown in Fig. 4.19b [89]. In this case, the grain sizes are estimated with an optical microscope, and a fit to the scaling law gives $R_S^G = 130\,\Omega$ and $\rho_{GB} = 1.4$ kΩ.μm. One useful consequence of using the scaling law is that it allows for an estimate of the average sheet resistance within the grains, R_S^D (for a good fit, it is best to have a range of grain sizes such that

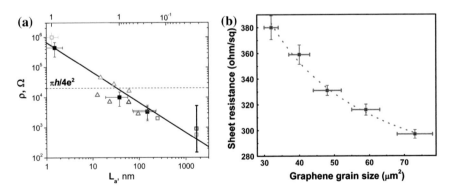

Fig. 4.19 Global measurements from scaling law. **a** Sheet resistance of Poly-G as a function of average grain size. Grain sizes were determined via Raman spectroscopy. Reproduced with permission [130], Copyright 2011, IOP Publishing. **b** Another example of the scaling behavior of Poly-G. The *dotted line* represents a fit to the scaling law described in the main text. Reproduced with permission [89]. Copyright 2012, Nature Publishing Group

$R_S^G < \rho_{GB}/l_G$ for the smallest grains and $R_S^G > \rho_{GB}/l_G$ for the largest grains). For example, based on the extracted values of R_S^G and ρ_{GB}, the GGBs begin to dominate the sheet resistance of these samples when the average grain size is less than $l_G = \rho_{GB}/R_S^G \approx 10\,\mu m$. This information can serve as a useful design parameter when considering large-scale applications of Poly-G.

4.2.6 Manipulation of GGBs with Functional Groups

4.2.6.1 Chemical Reactivity of GGBs

In addition to the general electrical transport properties of Poly-G, the chemical properties (reactivity, functionalization, etc.) of GGBs have been extensively discussed. For example, it has been shown theoretically that non-hexagonal atomic arrangements in graphene, such as the Stone-Wales defect, yield higher chemical reactivity than the ideal hexagonal structure [136–140], and this behavior has been extended to GGBs. A schematic representation is shown in Fig. 4.20a, where oxygen atoms preferentially attach to the non-hexagonal sites located in the GGBs. Selective oxidation of GGBs can be demonstrated by transferring CVD graphene to a mica substrate and heating the sample for 30 min at 500 °C. This process selectively burns away the GGBs [116], giving access to the grain morphology within the samples with AFM. A representative AFM image is given in Fig. 4.20b, where the dark lines indicate the location of the removed GGBs. This procedure not only provides a simple means of characterizing the grain morphology in the samples but also highlights the enhanced chemical reactivity of the GGBs.

UV treatment of Poly-G on a copper substrate can also reveal selective functionalization of the GGBs [89]. Under humid environment, O and OH radicals generated by the UV light preferentially attach to the GGBs, making the defects at the GGBs inert. This allows next incoming radicals to diffuse through large-pore heptagons and higher-order defects to eventually oxidize and expand the underlying copper substrate, as explained above. The degree of volume expansion can be engineered by controlling oxidation times, and the morphological changes around GGBs are easily identified by AFM and optical microscopy. The dark lines in Fig. 4.20c reveal the grain structure of the Poly-G. The grain structure is also revealed via Raman mapping of the sample, as shown in Fig. 4.20d–g. Figure 4.20d outlines the formation of a strong D-band associated with the GGBs after UV treatment. The D-peak also forms within the graphene grains, but its magnitude is much smaller, highlighting the higher chemical reactivity of the GGBs. Redshifts of the G and $2D$ (G') bands in the GGBs after UV treatment are attributed to strain induced by the oxidized copper below the GGBs. Figure 4.20e–g show that after UV treatment, spatial mappings of the D, G, and $2D$ peaks correlate well with the optical image of the GGBs. It should be noted that Raman mapping shows no evidence of the GGBs prior to UV treatment, indicating the strong influence by the oxidation of the GGBs.

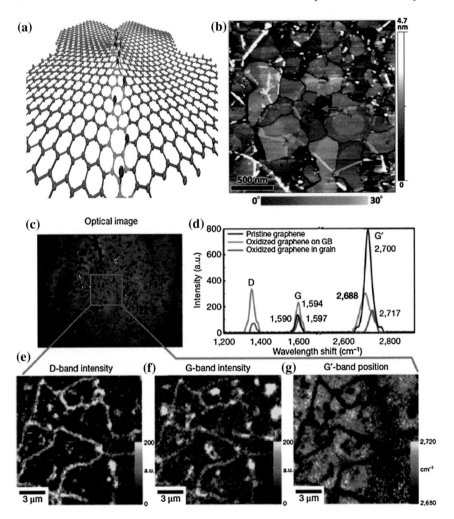

Fig. 4.20 Chemical reactivity of GGBs by experiments. **a** Representation of selective chemical functionalization of GGBs. **b** The location of GGBs can be imaged with AFM after burning them away at high temperature, which highlights their selective oxidation. Reproduced with permission [116]. Copyright 2011, AIP Publishing. **c** An optical image of Poly-G indicates the selective oxidation of an underlying copper substrate below the GGBs. **d** Raman spectroscopy indicates the strong oxidation at the GGBs after UV treatment. **e, f** Raman mapping indicates strong oxidation of the GGBs (*D*-band), as well as strain due to the expansion of the oxidized copper substrate below the GGBs (*G* and *G'* band shifts). Reproduced with permission [89]. Copyright 2012, Nature Publishing Group

The experimental demonstrations of the chemical reactivity of GGBs reported to date suggest that Poly-G may be a good material for the development of chemical sensors. For example, gas sensors based on pristine (single-grain) and Poly-G have yielded highly different responses to toluene and 1,2-dichlorobenzene, with the Poly-G sensor showing a response 50 times greater than that of pristine graphene [75]. This improvement in the sensitivity of the sensor is attributed to the increased reactivity of the GGBs and the enhanced impact that line defects have over point defects on transport features in two dimensions. This highlights the combined role that chemistry and charge transport play in the electrical properties of Poly-G.

4.2.6.2 Selective Functionalization of GGBs

As described above, GGBs are more chemically active than the graphene basal plane. However, selective functionalization of GGBs with an appropriate reactant is still an on-going area of research. Our main concern is a selective functionalization of GGBs, although defects inside grain could be functionalized as well. The whole graphene layer still retains metallicity with slightly increased sheet resistance. This is good contrast with heavily functionalized graphene oxide that leads to an insulator. Ozone is a good candidate for this purpose because it is inert with the graphene basal plane [141, 142]. Figures 4.21 and 4.22 shows measurements of the electrical reponse of the graphene basal plane and GGBs to ozone generated by UV exposure under an O2 environment. A four-probe device was fabricated on the merged region of two graphene domains (described in Fig. 4.16), as shown in Fig. 4.21. Series of Hall bar geometry ($5 \times 5\,\mu m^2$) was fabricated across through an expected GGB line as shown in Fig. 4.21a. The final device is shown in Fig. 4.21b, c after graphene parterning, metal deposition and lift-off process by e-beam lithography. After fabrication processes including graphene transfer and e-beam lithography, the GGBs and partial graphene basal plane are expected to be contaminated. Therefore, the sample was heat-treated at different conditions under vacuum (102 Torr). Physical adsorbates were simply removed at 150 °C for 1 h, and the transport characteristics

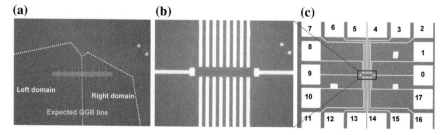

Fig. 4.21 Optical image of the four-probe device across a GGB. **a** E-beam lithography resist (PMMA) location at a merging region including a GGB. **b, c** A final device with Hall bar geometry at merging region of two graphene domains

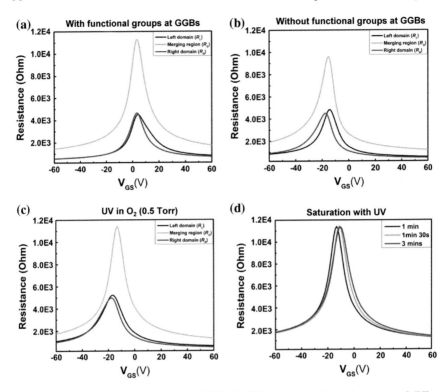

Fig. 4.22 O_2 Selective functionalization of GGBs by UV treatment under environment. **a, b** Effect of annealing at 250 °C in 3 h. Functional groups are removed from a GGB. **c, d** Effect of UV treatment under O_2 environment. The exclusive change of the inter-grain resistance indicates selective functionalization at the GGB. The UV treatment is saturated after 1 min of UV treatment

of the grains and the GGB were measured, as shown in Fig. 4.22a. Here, the black and blue lines represent the intra-grain resistances R_L and R_R, and the red line is the merging region resistance R_B. As expected, R_B is larger than R_L and R_R, due to the extra resistance contributed by the GGB. Next, the sample was further annealed at 250 °C for 3 h. Figure 4.22b shows that the resistance of the graphene basal plane was not changed, while the resistance across the GGB decreased significantly. This decrease in resistance implies that functional groups at the GGB were removed, as supported by the simulation results in the next section. The sample was then exposed to UV under an O2 environment (0.5*Torr*). The resistance across the GGB increased, while the resistance of the graphene basal plane was still unchanged, as shown in Fig. 4.22c. This strongly suggests that the GGBs are selectively functionalized by ozone generated by UV. This systematic series of measurements leads us to conclude that the GGBs can be selectively functionalized by ozone. This is a key step towards further biochemical modification of GGBs. We notice that the UV treatment is saturated after 1 min exposure. Longer time UV exposure doesnt increase the resistance at GGBs.

4.2.6.3 Effect of Functional Groups on Electrical Transport at GGBs by Simulation

As discussed above, the resistance at the GGBs can be modified by changing their functional groups. Proving this concept with a current measurement technique is a challenge because the chemical reaction occurs on the nanometer scale at the GGBs. Therefore, numerical simulation is a key strategy to understand this process. Several theoretical and numerical approaches have been employed to study charge transport across individual GGBs [21, 74, 143–146]. Here, an approach which allows the study of large-area Poly-G with a random distribution of GB orientations and morphologies is outlined. The Poly-G sample is created using molecular dynamics simulations that mimic the growth of CVD graphene [86], and its electrical properties are described with the TB formalism. To study transport, the time evolution of an electronic wave packet within the graphene sample is tracked [147]. The conductivity can then be calculated with the Kubo formula in Eq. (3.41). By assuming a wave packet that initially covers the entire sample, one can get a global picture of the scattering induced by GGBs. Once the conductivity is known, the sheet resistance is given by $R_S = 1/\sigma$. By doing this simulation for a range of average grain sizes, the GGB resistivity can be extracted using the scaling law described in Sect. 4.2.5.3. To include the effect of chemical functionalization, adsorbates are randomly attached to the GB atoms at different concentrations (as illustrated in Fig. 4.20a). Tight-binding parameters for describing hydrogen, hydroxyl, and epoxy groups have been taken from the literature [19, 20, 148].

Figure 4.23a, b shows a typical example of a 5-7 GGB functionalized by O and OH groups, respectively. The resistivity of the GGBs with different functional groups at various concentrations is extracted, as shown in Fig. 4.23c, where ρ_{GB} is plotted as a function of adsorbate coverage, defined as the number of adsorbates relative to the total number of GGB toms in the sample. For coverage greater than 100 %, the adsorbates are allowed to functionalize the carbon atoms next to the GGBs. For all types of adsorbates, ρ_{GB} increases with coverage, regardless of their type. However, it is also noted that ρ_{GB} is strongly adsorbate-dependent. For example, while both H and OH groups are chemisorbed to the top site of a single carbon atom, H groups have a stronger effect on transport through the GGBs than OH groups, with ρ_{GB} nearly 4 times larger at 200 % coverage. This difference can be ascribed to the electronic structure of each type of adsorbate. The simulations employ a resonant scattering model, where each adsorbate is characterized by an on-site energy ε_{ads} and a coupling to a single carbon atom γ_{ads}. The net effect of this model is to introduce an energy-dependent scattering potential [20], $V_{ads}(E) = \gamma_{ads}^2/(E - \varepsilon_{ads})$. Using parameters for H and OH taken from the literature [20, 147], this gives $V_H(E = 0) = -40\gamma_0$ and $V_{OH}(E = 0) = 1.8\gamma_0$. Since σ_{DC}, and hence R_S and ρ_{GB}, are calculated at the Dirac point, the H groups present a much stronger scattering potential than the OH groups. Calculations have also shown that H groups induce strongly localized states near the Dirac point, while OH adsorbates result in a more dispersive impurity band lying in the valence band of graphene [20]. Meanwhile, the O group chemisorbs in the bridge site by forming a pair with adjacent carbon atoms in the

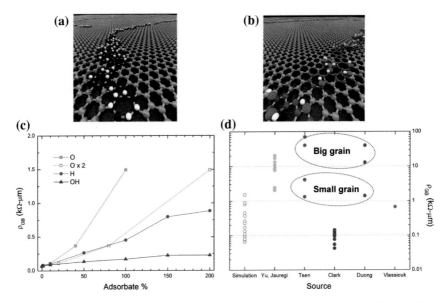

Fig. 4.23 Simulation of the effect of functional groups at GGBs. **a, b** Schematic of GGBs functionalized by *H* and *OH* groups, respectively. **c** Dependence of the resistivity of GGBs on functional groups with various concentrations. **d** Summary of experimental and simulated results for the resistivity of GGBs

graphene lattice (epoxide) [148]. The simulations clearly show that the resistance at GGBs with functional groups is much higher than that of pure GGBs. Figure 4.23d shows a summary of the values of ρ_{GB} derived from measurements compared to the simulation results [21, 88, 99, 100, 129, 148]. The solid symbols are from the electrical measurements described earlier In this section, and the open symbols are the numerical simulations. Here, most measurements give ρ_{GB} in the range of 1 to 10 kΩ.μm, except for one that gives values one to two orders of magnitude smaller [149]. This difference could be caused by the measurement technique, where ρ_{GB} was measured with four-probe STM under ultra-high vacuum, while the other groups fabricated physical contacts on their samples. This extra fabrication step could lead to additional contamination, increasing ρ_{GB}. Accordingly, the numerical simulations show that it is possible to bridge the gap between the various measurements by systematically increasing the amount of chemical functionalization of the GGBs. The situation becomes more complicated by several other parameters such as the structure and resistivity of the GGBs, as mentioned previously [21]. This is highlighted by the measurements labeled "small grain" and "large grain" in Fig. 4.23d, where growth conditions yielding large grain samples also tend to yield poorly connected and highly resistive GGBs [100]. Nevertheless, these results highlight the strong impact that chemical functionalization can have on the electrical properties of GGBs.

4.2.7   Challenges and Opportunities

The observation and characterization of GGBs at both atomic and macroscopic scale is mandatory to understand the transport properties and the related underlying physics and chemistry of Poly-G. As described in this Chapter, TEM and STM, combined with theory and simulation, can provide information at the atomic scale, with the related transport properties revealed with the assistance of STS. UV-treatment and liquid crystal coating, combined with optical microscopy, can provide information on both the GB distribution at the macro scale and the orientation of each domain, while macroscopic transport properties can be derived using the scaling law. With all these powerful methods available, one can envision their application to the engineering of GBs during graphene synthesis. For instance, ideal monocrystalline graphene could be obtained by designing seamless boundaries between coalescing graphene grains. With available large-area monocrystalline graphene, bilayer graphene with controlled stacking order can be constructed by aligned transfer techniques. The relative orientation of the layers can be identified by either low-energy electron diffraction or Raman spectroscopy. This opens a new research direction of bilayer graphene for designing vertical tunneling devices and planar switching devices.

A GB line is a 1D structure consisting of a series of pentagonal, hexagonal, and heptagonal carbon rings. It is possible to selectively functionalize as well as deposit designed materials only at the GGBs due to their higher chemical reactivity compared to ideal basal graphene. This implies that GGBs can be a good template for the synthesis of 1D materials. Atomic layer deposition, whose precursor is quite inert with the graphene basal plane, would be a good method for the synthesis of sub-nanometer 1D metals and semiconductors.

Another research direction to utilize GBs is to control their density to design sensors for detecting gases and molecules under different environmental conditions. As revealed by our numerical simulations and our experimental measurements, the transport properties of GBs can be strongly altered with chemical modifications of the GBs. Together with highly conductive graphene, electro-biochemical sensing devices with high sensitivity and selectivity could be designed.

Membrane science is another open research area. Although the ideal hexagonal graphene lattice impedes the diffusion of gases, defect sites such as heptagons, octagons, vacancies, and divacancies allow selective diffusion of limited gases and molecules, as mentioned above. This provides new opportunities to explore ultrafine membrane performance via the controlled engineering of GBs and point defects.

Although much progress has been made in the visualization and electrical characterization of GGBs from atomic scale to macro scale, issues still remain. The structure of GGBs is determined by the different orientations between merging domains, and the related physical and chemical properties are predicted to be strongly chirality-dependent. However, no electrical measurements have revealed such effects. The question is whether this originates from a device fabrication process which inevitably functionalizes GGBs, or if the native structure of GGBs is disordered, different from theoretical predictions.

GGBs also present challenges for the development of large scale graphene-based spintronic devices [150], and for harvesting the unique optical properties of graphene. For instance, GGBs introduce non-trivial local symmetry breaking which could significantly impact spin/pseudospin coupling and spin relaxation times, as well as the formation and propagation of plasmonic excitations. Similarly, the peculiar structure of interconnected GGBs could affect transport properties in high magnetic fields, such as the QHE. Overall, controlling the atomic structure of GGBs by CVD is a big challenge from a scientific point of view, but would be a huge step forward in the realization of next-generation technologies based on this material.

4.3 Impact of Graphene Polycrystallinity on the Performance of Graphene Field-Effect Transistors

4.3.1 Introduction

In the effort to successfully realize next-generation technologies based on graphene field-effect transistors (GFETs), theory and device modeling will play a crucial role. Specifically, it is important to develop models that can accurately describe both the electrostatics and the current-voltage (I-V) characteristics of graphene-based electronic devices [151–154]. This capability will enable device design optimization and performance projections, will permit benchmarking of graphene-based technology against existing ones [53, 155], and will help to explore the feasibility of analog/RF circuits based on graphene [156–158]. Ultimately, graphene-based devices could provide new or improved functionality with respect to existing technologies, such as those based on silicon or III-V materials.

The CVD technique for growing wafer-scale graphene on metallic substrates [64, 159–161] produces a polycrystalline pattern. This is because the growth of graphene is simultaneously initiated at different nucleation sites, leading to samples with randomly distributed grains of varying lattice orientations [72]. It has recently been predicted that the electronic properties of Poly-G differ from those of pristine graphene, where the mobility scales linearly with the average grain size [21]. Based on these results, we report on how the electronic properties of Poly-G impact the behavior of graphene-based devices. Specifically, we concentrate our study on the effect that Poly-G has on the gate electrostatics and I-V characteristics of GFETs. We find that the source-drain current and the transconductance are proportional to the average grain size, indicating that these quantities are hampered by the presence of GBs in the Poly-G. However, our simulations also show that current saturation is improved by the presence of GBs, and the intrinsic gain is insensitive to the grain size. These results indicate that GBs play a complex role in the behavior of graphene-based electronics, and their importance depends on the application of the device.

4.3.2 Poly-G Effect on the Gate Electrostatics and I-V Characteristics of GFETs

The starting point of our study is the characterization of a large-area model of disordered Poly-G samples, containing hundreds of thousands atoms and described by varying grain misorientation angles, realistic carbon ring statistics, and unrestricted GB structures, based on the method reported in Ref. [118]. To calculate the electronic and transport properties, we used a TB Hamiltonian and an efficient quantum transport method [41, 43], which is particularly well-suited for large samples of disordered low-dimensional systems. The transport calculations were based on a real-space order-N quantum wave packet evolution approach, which allowed us to compute the Kubo-Greenwood conductivity (Eq. 3.41). With this quantity, the charge carrier mobility can be estimated as $\mu(E) = \sigma(E)/q^*Q_c(E)$, where Q_c is the 2D charge density in the graphene. It should be noted that we assume the carrier mobility is not limited by the substrate, that is, we do not consider additional scattering due to charge traps or surface phonons in the insulator that could further degrade the carrier mobility [162]. Thus, our results represent an upper bound on the performance metrics of the GFETs that we are studying.

In this work, we focus on a dual-gate GFET as the one depicted in Fig. 4.24. This transistor is based on a metal/oxide/Poly-G/oxide/semiconductor structure where an external electric field modulates the mobile carrier density in the Poly-G layer. The electrostatics of this dual gate structure can be understood with an application of Gauss law

$$Q_c = C_t(V_{gs}^* - V_c) + C_b(V_{bs}^* - V_c) \qquad (4.7)$$

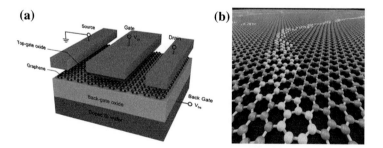

(a)   **(b)**

Fig. 4.24 **a** Schematic of the dual-gate GFET, consisting of a poly-G channel on *top* of an insulator layer, which is grown on a heavily-doped Si wafer acting as the back gate. An artistic view of the patchwork of coalescing graphene grains of varying lattice orientations and size is shown in (**b**). The source and drain electrodes contact the poly-G channel from the *top* and are assumed to be ohmic. The source is grounded and considered the reference potential in the device. The electrostatic modulation of the carrier concentration in graphene is achieved via a top-gate stack consisting of the gate dielectric and the gate metal

where $Q_c = q(p - n)$ is the net mobile charge density in the graphene channel, C_t and C_b are the geometrical top and bottom oxide capacitances, and V_{gs}^* and V_{bs}^* are the effective top and bottom gate-source voltages, respectively. Here, $V_{gs}^* = V_{gs} - V_{gs0}$ and $V_{bs}^* = V_{bs} - V_{bs0}$, where V_{gs0} and V_{bs0} are quantities that comprise the work function differences between each gate and the graphene channel, charged interface states at the graphene/oxide interfaces, and possible doping of the graphene. The graphene charge density can be determined numerically using the procedure

$$Q_c(V_c) = q \int_{-\infty}^{0} DOS_{p-G}(E)f(qV_c - E)dE - q \int_{0}^{\infty} DOS_{p-G}(E)f(E - qV_c)dE \quad (4.8)$$

where $DOS_{p-G}(E)$ has been calculated with the procedure outlined in Ref. [21]. The potential V_c represents the voltage drop across the graphene layer, and is related to the quantum capacitance C_q of the Poly-G by $C_q = -dQ_c/dV_c$. When the entire length of the transistor is considered, the effective gate voltages can be written as $V_{gs}^* = V_{gs} - V_{gs0} - V(x)$ and $V_{bs}^* = V_{bs} - V_{bs0} - V(x)$, where $V(x)$ (the so-called quasi-Fermi level) represents the potential along the graphene channel. The boundary conditions that should be satisfied are $V(0) = 0$ at the source and $V(L) = V_{ds}$ at the drain.

To model the drain current, we employ a drift-diffusion model with the form $I_{ds} = -W|Q_c(x)|v(x)$, where W is the gate width, $Q_c(x)$ is the free carrier sheet density in the channel at position x, and $v(x)$ is the carrier drift velocity. The latter is related to the transverse electric field E as $v = \mu E$, so no velocity saturation effect has been included in this model. The low-field carrier mobility $\mu(Q_c)$ is density-dependent and calculated via the procedure of Ref. [21]. After applying $E = -dV(x)/dx$, including the above expression for v, and integrating the resulting equation over the device length, the source-drain current becomes

$$I_{ds} = \frac{W}{L} \int_{0}^{V_{ds}} \mu|Q_c|dV. \quad (4.9)$$

In order to calculate I_{ds}, the integral in Eq. (4.9) is solved using V_c as the integration variable and subsequently expressing μ and Q_c as functions of V_c, based on the mapping given by Eq. (4.8). This gives

$$I_{ds} = \frac{W}{L} \int_{V_{cs}}^{V_{cd}} \mu(V_c)|Q_c(V_c)|\frac{dV}{dV_c}dVc \quad (4.10)$$

where V_c is obtained by self-consistently solving Eqs. (4.7) and (4.8). The channel potential at the source is determined as $V_{cs} = V_c(V = 0)$ and the channel potential at the drain is determined as $V_{cd} = V_c(V = V_{ds})$. Finally, Eq. (4.7) allows us to evaluate the derivative appearing in Eq. (4.10), namely, $\frac{dV}{dV_c} = -1 + \frac{C_q}{C_t + C_b}$, which should be determined numerically as a function of the integration variable V_c.

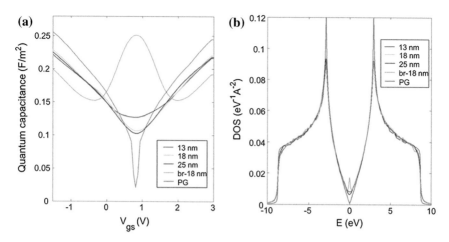

Fig. 4.25 Quantum capacitance (**a**) and density of states (**b**) of Poly-G considering different average grain sizes. The pristine graphene case has also been plotted for the sake of comparison

Next, we apply the multi-scale model to the GFET shown in Fig. 4.24. It consists of a dual-gate structure with $L = 10\,\mu m$ and $W = 5\,\mu m$. The top and bottom gate insulators are hafnium oxide and silicon oxide with thicknesses of 4 and 300 nm, respectively. For the active channel, we considered poly-G with different average grain sizes together with the simple pristine graphene case, which serves as a convenient reference for comparison. For this study, we created samples with three different average grain sizes (average diameter $\langle d \rangle \approx 13$, 18, and 25.5 nm) and uniform grain size distributions. The atomic structure at the GBs consists predominantly of five- and seven-member carbon rings and assumes meandering shapes similar to the experimentally observed ones. We also created one sample with $\langle d \rangle \approx 18$ nm and "broken" (poorly connected) boundaries ("br-18 nm"). The quantum capacitance (C_q) of each sample is presented in Fig. 4.25a, which reflects the structure of the DOS, shown in Fig. 4.25b. An enhanced density of zero-energy modes around the charge neutrality point (CNP) can be observed, which arises locally from the atomic configurations of the GBs, giving rise to a finite C_q. A zero C_q would correspond to ideal gate efficiency, meaning that the gate voltage would have 100 % control over the position of the graphene Fermi level. Away from the CNP, both C_q and the DOS of the analyzed structures look very similar. For the poorly connected sample "br-18 nm", a peak is observed around the CNP because of a higher density of midgap states, resulting in a negative differential C_q.

Figure 4.26a shows the transfer characteristics of the GFET under consideration for different grain sizes. The low-field carrier mobility was calculated from the Kubo-Greenwood conductivity as $\mu(E) = \sigma(E)/q^*Q_c(E)$, and has been plotted as a function of Q_c in Fig. 4.26b. The mobility corresponding to a grain size of $1\,\mu m$ was estimated from the mobility at 25.5 nm with a simple scaling law [21], $\mu_{1\mu m}(Q_c) = (1\,\mu m/25.5\,nm)\mu_{25.5\,nm}(Q_c)$. The resulting I-V characteristics exhibit

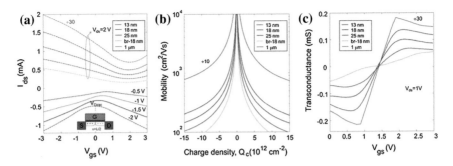

Fig. 4.26 Transfer characteristics (**a**) and transconductance (**c**) of the graphene field-effect transistor considering different samples of Poly-G as the active channel. **b** Estimated low-field carrier mobility as a function of the carrier density for each of the samples

the expected V-like shape with an ON-OFF current ratio in the range of 2–4, and one can see that the source-drain current is proportional to the average grain size. This is due to the scaling of the mobility with grain size, as shown in Fig. 4.26b. In Fig. 4.26c, we plot the transconductance of the GFET, defined as $g_m = dI_{ds}/dV_{gs}$, which is a key parameter in determining the transistor voltage gain or the maximum operation frequency. It appears that small grain sizes are detrimental to this factor. The reason behind such a degradation is the combination of two factors as the grain size is reduced: (a) an increase in C_q at low carrier densities (Fig. 4.25a), which is related with the increase in the DOS near the CNP (Fig. 4.25b) and leads to reduced gate efficiency; and (b) the reduction of the low-field carrier mobility (Fig. 4.26b) because of scattering due to the disordered atomic structure of the GBs. Figure 4.26b indicates that the mobility is proportional to the average grain size of the Poly-G; a higher density of GBs results in more scattering and a lower mobility. The scattering effect of the GBs has been further quantified in Ref. [21], which shows the scaling of the conductivity and the mean free path of the Poly-G for different grain sizes. For example, the sample with 25.5-nm grains has a mean free path of 10 nm near the Dirac point, compared with 5 nm for the sample with 13-nm grains.

In Fig. 4.27a, we plot the GFET output characteristics for different grain sizes and gate biases. The output characteristic exhibits an initial linear region dominated by hole transport (p-type channel), followed by a weak saturation region. The onset of saturation ($V_{sd,sat}$) happens when the channel becomes pinched off at the drain side. A further increase in V_{sd} drives the transistor towards the second linear region, characterized by a channel with a mixed p- and n-type behavior. Interestingly, a reduction of the grain size improves the current saturation, which can be seen in a plot of the output conductance (Fig. 4.27b), defined as $g_d = VVdI_{ds}/dV_{ds}$. Here, the minimum of g_d is much flatter and broader for smaller grain sizes. Both g_m and g_d determine the intrinsic gain $A_v = g_m/g_d$, which is a key figure of merit in analog or RF applications. Our simulations demonstrate that A_v is insensitive to the grain size (Fig. 4.28), because an increase in g_m is almost exactly compensated by a similar increase in g_d. This suggests that polycrystallinity is not a limiting factor in analog/RF

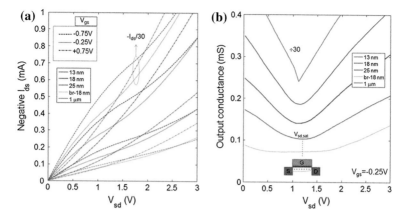

Fig. 4.27 Output characteristics (**a**) and output conductance (**b**) of the graphene field-effect transistor considering different samples of Poly-G as the active channel

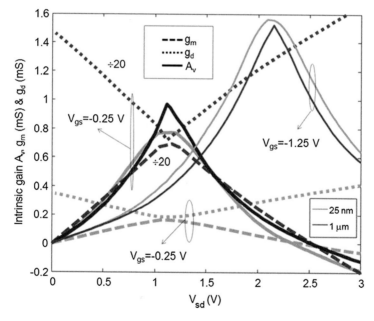

Fig. 4.28 Intrinsic gain as a function of the drain voltage. The transconductance and output conductance are also plotted at $V_{gs} = 0.25\,\text{V}$

devices whose performance depends on the intrinsic gain. However, there are other performance metrics, such as the intrinsic cutoff (f_T) and maximum frequencies (f_{max}), which are severely degraded by the presence of GBs. To demonstrate this, we have calculated both f_1 and f_{max} for the device under consideration, but assuming a channel length of 100 nm. The cutoff frequency is given by $f_T \approx g_m/2\pi C_{gs}$, where

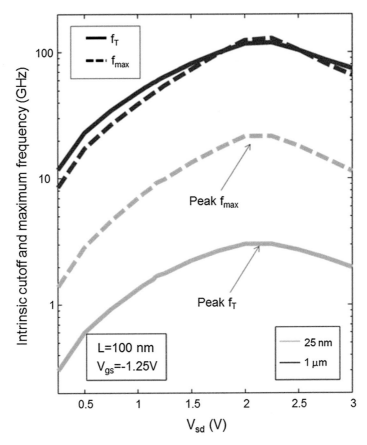

Fig. 4.29 Intrinsic maximum and cutoff frequency for the simulated transistor assuming a channel length of 100 nm

C_{gs} is the gate-to-source capacitance. 12 Given that the geometrical capacitance C_t is much smaller than the quantum capacitance C_q, $C_{gs} \cong C_t$. The maximum frequency is given by $f_{max} \approx g_m/(4\pi C_{gs}\sqrt{g_d(R_S + R_G)})$, where R_S and R_G are the source and gate resistances, respectively [155]. Here, we have assumed state of the art values, such as [163] $R_S \sim 100\,\Omega.\mu m$ and $R_G \sim 6\,\Omega$. As shown in Fig. 4.29, f_{max} and f_T are degraded by one and two orders of magnitude, respectively, when the average grain size decreases from 1 μm to nm.

Realistic GFETs are limited in performance by interaction with the substrate and top gates. Comparing with the extracted mobility from some reported state-of-the-art devices [164], our calculations, which represent the limiting case of uncovered graphene, overestimate the mobility of these devices by $\sim 10\times$. As a consequence, g_m, g_d, and f_T should be reduced by that amount when considering substrate and top gate effects. Meanwhile, A_v is expected to remain constant and f_{max} is expected to be reduced by $\sim 3\times$. The mentioned $\sim 10\times$ factor of mobility reduction could be made significantly smaller by using an appropriate substrate, such as diamond-like carbon [162] (DLC), which helps to minimize interaction with the substrate.

In conclusion, we have developed a drift-diffusion transport model for the GFET, based on a detailed description of electronic transport in poly-G. This model allows us to determine how a graphene sample's polycrystallinity alters the electronic transport in GFETs, enabling the prediction and optimization of various figures of merit for these devices. We have found that the presence of GBs produces a severe degradation of both the maximum frequency and the cutoff frequency, while the intrinsic gain remains insensitive to the presence of GBs. Overall, polycrystallinity is predicted to be an undesirable trait in GFETs targeting analog or RF applications.

4.4 Transport Properties of Amorphous Graphene

4.4.1 Introduction

The physics of disordered graphene is at the heart of many fascinating properties such as Klein tunneling, WAL or anomalous QHE (see reviews [43, 165]). The precise understanding of individual defects on electronic and transport properties of graphene is currently of great interest [166]. For instance, graphene samples obtained by large-scale production methods display a huge quantity of structural imperfections and defects which jeopardize the robustness of the otherwise exceptionally high charge mobilities of their pristine counterparts [8]. Indeed, the lattice mismatch-induced strain between graphene and the underlying substrate generates Poly-G with GBs which strongly impact on transport properties [143] (See Sect. 4.2). However, despite the large amount of disorder, such graphene flakes usually maintain a finite conductivity down to very low temperatures (when deposited onto oxide substrates) owing to electron-hole puddles (charge inohomogeneities fluctuations)-induced percolation effects which limit localization phenomena [9]. The predicted Anderson localization in two-dimensional disordered graphene has been hard to measure in non intentionally damaged graphene, in contrast to chemically modified graphene [167, 168]. In a recent experiment, it was however possible to screen out electron-holes puddles using sandwiched graphene in between two boron-nitride layers, together with an additional graphene control layer [12]. As a result of puddles screening, a large increase of the resistivity was obtained at the Dirac point, evidencing an onset of the Anderson localization regime.

Beyond individual defects and polycrystallinity, a higher level of disorder can be induced on graphene to the point of obtaining two-dimensional amorphous networks composed of sp^2 hybridized carbon atoms. Such networks contain rings other than hexagons in a disordered arrangement. The average ring size is six according to Euler's theorem, allowing such a system to exist as a flat 2D structure. Experimentally, such amorphous two-dimensional lattices have been obtained in electron-beam irradiation experiments [86], and directly visualized by high resolution electron trans mission microscopy. Previously, indirect evidence for the formation of an amorphous network was obtained by Raman spectroscopy in samples subject to electron-beam irradiation [169], ozone exposure [170] and ion irradiation [171]. In all these cases, an evolution from polycrystalline to amorphous structures was observed upon increase of the damage treatment. In [171], further evidence of the formation of an amorphous network was obtained through transport measurements. These indicate the transition from a WL regime in the polycrystalline samples to variable range hopping transport in the strongly localized regime for amorphous samples, as evidenced by the temperature dependence of the conductivity. Localization lengths were estimated to be of the range 0.1–10 nm in the amorphous samples, depending on the degree of amorphization. From the theoretical side, models of the amorphous network have been proposed using stochastic quenching methods [172], and molecular dynamics [46, 173, 174]. Electronic structure calculations show that the amorphization yields a large increase of the density of states at and in the environment of the charge neutrality point [172–174]. Despite the expected reduction of the conduction properties due to strong localization effects, Holmström et al. [173] suggest that disorder could enhance metallicity in amorphized samples, in contrast with the experimental evidence.

Here, we explore the transport properties of two-dimensional sp^2 lattices with massive amount of topological disorder, encoded in a geometrical mixture of hexagons with pentagon and heptagon rings with a given ring statistics. The calculations are done using two approaches: a Kubo formulation in which the conductivity of bulk 2D amorphous graphene lattices was determined, and a Landauer-Büttiker formulation where the conductance of ribbons of amorphous graphene contacted to semi-infinite pristine graphene electrodes was calculated. Both approaches lead to similar findings. Depending on the ratio between odd versus even-membered rings, a transition form a graphene-like electronic structure to a totally amorphous and smooth electronic distribution of states is obtained. The stronger the departure from the pristine graphene, the more insulating is the corresponding lattice, which transforms into a strong Anderson insulator with elastic mean free paths below one nanometer and very short localization length all over the whole electronic spectrum. Those structures are therefore inefficient to carry any sizable current, and are therefore useless for any practical electronic applications such as touch screens displays or conducting electrodes, but interesting for scrutinizing localization phenomena in low dimensional materials.

4.4.2 Models of Amorphous Graphene

Amorphous models of graphene are prepared using the Wooten-Winer-Weaire (WWW) method [175, 176], introducing Stone-Wales defects [177] into the perfect honeycomb lattice. To generate the structures, periodic boundary conditions are imposed and the entire network was relaxed with the Keating-like potential [172, 178]. Pieces of two different networks are shown in Fig. 4.30a, b. The samples con-

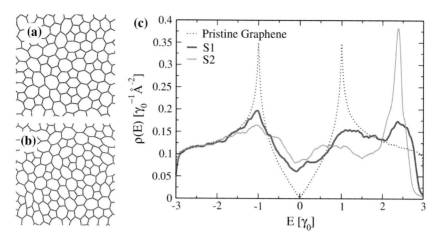

Fig. 4.30 **a** and **b** show details of amorphous graphene samples *S1* and *S2*, respectively, used to compute the conductivity with the Kubo approach. **c** Total density of states of the two amorphous samples. The pristine crystalline graphene case (*dashed lines*) is also shown for comparison

Table 4.2 Comparison of sample specifications

	S1	S2
Number of atoms	10032	101640
Percent. of n-membered rings ($n = 5/6/7$)	24/52/24	44/12/44
$\langle n^2 \rangle - \langle n \rangle^2$	0.47	0.88
RMS deviation of bond angles	11.02°	18.09°
RMS deviation of bond lengths	0.044 Å	0.060 Å
Fermi energy (γ_0)	0.03	0.05

tain 10032 and 101640 atoms, respectively, all of them with three-fold coordination as the honeycomb lattice, but topologically distinct. Samples 1 and 2 are characterized by a number of parameters given in Table 4.2. For Sample 1, 24 % of the elementary rings are pentagons, 52 % hexagons and 24 % heptagons, while sample 2 has a larger share of odd-membered rings. In both samples, the number of heptagons is the same as that of pentagons, according to Euler's theorem, and these systems can exist without an overall curvature as flat 2D structures with some distortions of bond lengths and angles, although may pucker under some circumstances. We will only be concerned with the planar structures here.

For the calculation of the Landauer-Büttiker conductance, we set up models in which an amorphous ribbon is contacted by two pristine graphene electrodes at a distance L. Models with different ribbon length of the amorphous contact are built to study the dependence of the conductance on the distance between electrodes. The models are periodic in the direction perpendicular to the ribbon, with a periodicity of $W = 11.4$ nm, and have the same ring statistics as the bulk sample 1 described above.

4.4.3 Electronic Properties

The electronic and transport properties of these disordered lattices are investigated using π-π^* orthogonal TB model with nearest neighbors hopping γ_0 and zero onsite energies. No variation of the hopping elements with disorder is included in the model as bond-length variation does not exceed a few percent (cf. Table 4.2); all dependence on disorder stems from the ring statistics which is the dominating effect. Figure 4.30c shows the density of states (DOS) of the two disordered samples, together with the pristine case (dashed line) for comparison. Sample 1, which keeps 52 % of hexagonal rings, displays several noticeable features, similar to those found in previous studies [172, 173]. First, the DOS at the charge neutrality point is found to be increased by a large amount. Additionally, the electron-hole symmetry of the band structure is broken due to the presence of odd-membered rings and the resonant states that these induce [32]. The hole part of the spectrum is still reminiscent of the graphene DOS, with a smoothened peak at the van Hove singularity while in the electron part a second maximum appears close to the upper conduction band edge. By reducing further the ratio of even versus odd-membered rings (Sample 2), the second maximum develops to a strong peak at about $E = 2.5\gamma_0$ while spectral weight at $E = 3\gamma_0$ is suppressed. The redistribution of DOS at the upper conduction band edge is a signature of odd-membered rings and its strength with increasing number of such rings relates the statistical distribution of rings with the DOS features.

Transport Methodology. To explore quantum transport in these topologically disordered graphene bulk samples, we employ a real-space order-N quantum wavepacket evolution approach in Chap. 3 to compute the Kubo-Greenwood conductivity [179]. The conductance of amorphous stripes (ribbons) contacted to graphene electrodes is computed using the Landauer-Büttiker approach [180]:

$$G(E) = G_0 T(E) = \frac{2e^2}{h}\mathrm{Tr}\left[t^\dagger t\right] \qquad (4.11)$$

where $T(E)$ and $t(E)$ are the transmission probability and transmission matrix, respectively, which can be computed from the Green's function $G(E)$ in the contact region and the broadening $\Gamma(E)$ of the states due to the interaction with the left and right electrodes. We calculate the conductance of the ribbon, which is infinite and periodic in the direction parallel to the interface between the pristine graphene electrodes and the amorphous ribbons. Despite the very large periodicity of our models, we perform a thorough sampling of the $k_\parallel$-points in that direction [181, 182], to obtain the appropriate V-shaped conductance of graphene in the thermodynamic limit. G is given per supercell of periodicity $W = 11.4$ nm. Note that conductivity and conductance are related though $\sigma = \frac{L}{W}G$.

Mean Free Path, Conductivity and Localization Effects. Figure 4.31 shows time dependence of the normalized diffusion coefficient $D(t)/D_{\max}$ for two chosen energies, for the two bulk samples. For energy $E = -2\gamma_0$, it is found to increase ballistically at short time, but then saturates typically after 0.1 ps. This saturation allows to

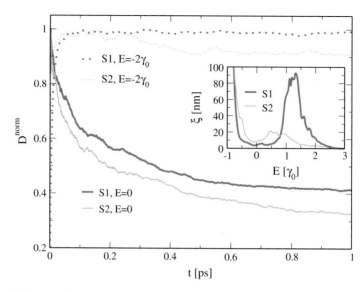

Fig. 4.31 Normalized time-dependent diffusion coefficients for two selected energies for both samples *S1* and *S2*. *Inset* localization lengths as a function of the carrier energy

extract the corresponding mean free paths $\ell_e(E)$. Localization effects, manifested in a decay of the diffusion coefficient with time, are apparent for the lines corresponding to the charge neutrality point, but are less clear for $E = -2\gamma_0$.

The elastic mean free path and the semiclassical conductivities are shown in Fig. 4.32, as obtained from the maximum of the diffusion coefficient. A striking feature is the very low value of the mean free path ℓ_e below 0.5 nm for the energy window around the Fermi level, in which the DOS departs from that of the pristine graphene structure. For negative energies (holes) far from the charge neutrality point, a considerable increase of more than one order of magnitude in the mean free paths is observed. The increase occurs for smaller binding energies for sample 1 than for sample 2, in good correlation with the changes observed in the DOS (which, around the van Hove singularity, deviates from the pristine graphene one more strongly for sample 2).

The semiclassical conductivities show a minimum value at the Fermi level close to $\sigma_{sc}^{min} = 4e^2/\pi h$, in agreement with the values for graphene in the presence of disorder induced by impurities or scatterers [28, 29]. We note, however, that the conductivity remains nearly constant at that value for an energy range of several eV around the Fermi level. This indicates that transport is strongly degraded in the amorphous network compared to pristine graphene, in which the conductivity increases rapidly away from the Fermi level. The charge mobility, $\mu(E) = \sigma_{sc}(E)/en(E)$, with $n(E)$ being the carrier density, is found to be of the order of 10 cm^{2}V^{-1}s^{-1} for $n = 10^{11}$–10^{12}cm^{-2}, which is orders of magnitudes lower than those usually measured in graphene samples [11]. Such low conductivity and mobility values should be measured at room temperature, where the semiclassical approximation is expected to hold.

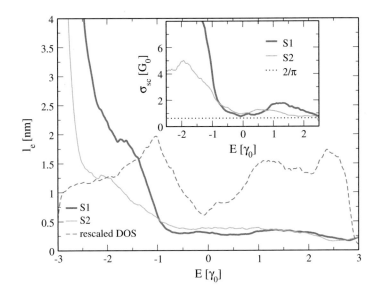

Fig. 4.32 Elastic mean free path versus energy for the two samples. DOS of sample *S1* is also shown for comparison in rescaled units. *Inset* semiclassical conductivity of corresponding lattices

The very short mean free paths obtained indicate a further significant contribution of quantum interferences turning the system to a weak and strong insulating system as the temperature drops. Interference effects are evidenced by the time-dependent decay of the diffusion coefficient $D(t)/D_{\max}$. Based on the scaling theory of localization [183], an estimate of the localization length of electronic states can be extracted from the semiclassical parameters by $\xi(E) = \ell_e(E) \exp(\pi \hbar \sigma_{sc}(E)/2e^2)$. The results are shown in Fig. 4.31 (inset). The amorphous samples are extremely poor conductors, with localization lengths as low as $\xi \sim 5\text{--}10$ nm over a large energy window around the charge neutrality point.

To further confirm the localization lengths estimated using scaling theory, we compute explicitly the conductance of the amorphous graphene ribbons contacted with pristine graphene electrodes, as a function of the ribbon length L. Figure 4.33 shows the conductance curves for two ribbons of 1.6 and 8.6 nm, respectively, compared to that of a graphene contact with the same lateral size in the supercell (11.4 nm). It is clear that the conductance of the amorphous samples is greatly reduced with respect to that of graphene, and that the reduction is more pronounced as the length of the amorphous ribbon becomes larger. Also, while the conductance for the ribbon with the smallest length is relatively smooth, it becomes more noisy as the ribbon becomes longer. This reflects the transition from a diffusive system, in which the ribbon is longer than the mean free path, but shorter than the localization length, to a strongly localized one in which the localization length is shorter than the ribbon length.

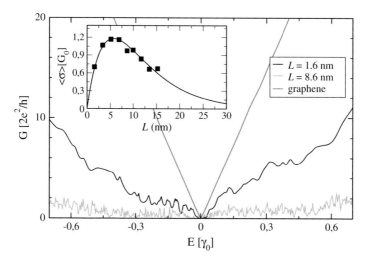

Fig. 4.33 Landauer-Büttiker conductance (for $W = 11.4$ nm) of two amorphous ribbons contacted to graphene electrodes with $L = 1.6$ and 8.6 nm, respectively. The conductance of a pristine graphene contact with the same lateral size (11.4 nm) is shown for comparison. The *inset* shows the dependence of the conductivity on the ribbon size L; symbols: calculated points; *line* fit to $\sigma(L) \sim \frac{L}{W} e^{-L/\xi}$

From the variation of the Landauer-Büttiker conductance with size L, we can extract reliable values of the localization lengths, as in the Anderson regime the conductance should decay as $G(L) \sim e^{-L/\xi}$. The inset in Fig. 4.33 shows the value of the conductivity, obtained from the conductance, for each size from 1.6 to 15.3 nm, averaged over an energy window of $1.5\gamma_0$ around the Fermi energy. A fit of the results to $\sigma(L) \sim \frac{L}{W} e^{-L/\xi}$ yields a value of $\xi = 5.8$ nm. This value is consistent with that obtained above using scaling theory, for energies close to the Fermi level, and confirms that, in these amorphous structures, strong localization effects should occur at low temperatures at distances of less than 10 nm. These estimates are in good agreement with the experimental results from transport measurements by Zhou et al. [171], which show values in the range between 0.1 and 10 nm for samples amorphized by ion radiation.

In conclusion, we have shown that amorphous graphene is a strong Anderson insulator. The increase of the density of states close to the charge neutrality point is concomitant with marked quantum interferences which inhibit current flow at low temperature. Very short mean free paths and localization lengths are predicted, in line with recent experimental evidence in graphene under heavy ion irradiation damage [171].

References

1. M.I. Katsnelson, K.S. Novoselov, A.K. Geim, Nat. Phys. **2**, 620 (2006)
2. A.F. Young, P. Kim, Nat. Phys. **5**, 222 (2009)
3. E. McCann, K. Kechedzhi, V.I. Falko, H. Suzuura, T. Ando, B.L. Altshuler, Phys. Rev. Lett. **97**, 146805 (2006)
4. F.V. Tikhonenko, A.A. Kozikov, A.K. Savchenko, R.V. Gorbachev, Phys. Rev. Lett. **103**, 226801 (2009)
5. K.S. Novoselov, A.K. Geim, S.V. Morozov, D. Jiang, M.I. Katsnelson, I.V. Grigorieva, S.V. Dubonos, A.A. Firsov, Nature (London) **438**, 197 (2005)
6. Y. Zhang, Y.-W. Tan, H.L. Stormer, P. Kim, Nature (London) **438**, 201 (2005)
7. A.H. Castro Neto, F. Guinea, N.M.R. Peres, K.S. Novoselov, A.K. Geim, Rev. Mod. Phys. **81**, 109 (2009)
8. K.S. Novoselov, Rev. Mod. Phys. **83**, 837 (2011)
9. S.D. Sarma, S. Adam, E.H. Hwang, E. Rossi, Rev. Mod. Phys. **83**, 407 (2011)
10. S.V. Morozov, K.S. Novoselov, M.I. Katsnelson, F. Schedin, D.C. Elias, J.A. Jaszczak, A.K. Geim, Phys. Rev. Lett. **100**, 016602 (2008)
11. Y.W. Tan, Y. Zhang, K. Bolotin, Y. Zhao, S. Adam, E.H. Hwang, S.D. Sarma, H.L. Stormer, P. Kim, Phys. Rev. Lett **99**, 246803 (2007)
12. L.A. Ponomarenko, A.K. Geim, A.A. Zhukov, R. Jalil, S.V. Morozov, K.S. Novoselov, I.V. Grigorieva, E.H. Hill, V.V. Cheianov, V.I. Falko, K. Watanabe, T. Taniguchi, R.V. Gorbachev, Nat. Phys. **7**, 958 (2011)
13. F. Evers, A.D. Mirlin, Rev. Mod. Phys. **80**, 1355 (2008)
14. S.D. Sarma, E.H. Hwang, Q. Li, Phys. Rev. B **85**, 195451 (2012)
15. M.M. Ugeda, I. Brihuega, F. Guinea, J.M. Gomez-Rodrguez, Phys. Rev. Lett. **104**, 096804 (2010)
16. V.M. Pereira, F. Guinea, J.M.B. Lopes dos Santos, N.M.R. Peres, A.H.C. Neto, Phys. Rev. Lett. **96**, 036801 (2006)
17. V.M. Pereira, J.M.B. Lopes dos Santos, N.M.R. Peres, A.H.C. Neto, Phys. Rev. B **77**, 115109 (2008)
18. N.M.R. Peres, F. Guinea, A.H.C. Neto, Phys. Rev. B **73**, 125411 (2006)
19. J.P. Robinson, H. Schomerus, L. Oroszlany, V.I. Falko, Phys. Rev. Lett. **101**, 196803 (2008)
20. T.O. Wehling, S. Yuan, A.I. Lichtenstein, A.K. Geim, M.I. Katsnelson, Phys. Rev. Lett. **105**, 056802 (2010)
21. D. Van Tuan, J. Kotakoski, T. Louvet, F. Ortmann, J.C. Meyer, S. Roche, Nano Lett. **13**, 1730 (2013)
22. J. Lahiri, Y. Lin, P. Bozkurt, I.I. Oleynik, M. Batzill, Nat. Nanotechnol. **5**, 326 (2010)
23. P.M. Ostrovsky, I.V. Gornyi, A.D. Mirlin, Phys. Rev. B **74**, 235443 (2006)
24. P.M. Ostrovsky, M. Titov, S. Bera, I.V. Gornyi, A.D. Mirlin, Phys. Rev. Lett. **105**, 266803 (2010)
25. J. Tworzydlo, B. Trauzettel, M. Titov, A. Rycerz,C. W., J. Beenakker, Phys. Rev. Lett. **96**, 246802 (2006)
26. F. Miao, S. Wijeratne, Y. Zhang, U.C. Coskun, W. Bao, C.N. Lau, Science **317**, 1530 (2007)
27. S. Yuan, H. De Raedt, M.I. Katsnelson, Phys. Rev. B **82**, 115448 (2010)
28. N.H. Shon, T. Ando, J. Phys. Soc. Jpn. **67**, 2421 (1998)
29. K. Nomura, A.H. MacDonald, Phys. Rev. Lett. **98**, 076602 (2007)
30. T. Stauber, N.M.R. Peres, F. Guinea, Phys. Rev. B **76**, 205423 (2007)
31. W. Li, W. Zhu, Q.W. Shi, X.R. Wang, X.P. Wang, J.L. Yang, J.G. Hou, Phys. Rev. B **85**, 073407 (2012)
32. A. Lherbier, S.M.M. Dubois, X. Declerck, S. Roche, Y.M. Niquet, J.C. Charlier, Phys. Rev. Lett. **106**, 046803 (2011)
33. T.M. Radchenko, A.A. Shylau, I.V. Zozoulenko, Phys. Rev. B **86**, 035418 (2012)
34. S.-Z. Liang, J.O. Sofo, Phys. Rev. Lett. **109**, 256601 (2012)

35. J.-H. Chen, W.G. Cullen, C. Jang, M.S. Fuhrer, E.D. Williams, Phys. Rev. Lett. **102**, 236805 (2009)
36. J. Yan, M.S. Fuhrer, Phys. Rev. Lett. **107**, 206601 (2011)
37. L. Zhao, R. He, K.T. Rim, T. Schiros, K.S. Kim, H. Zhou, C. Gutierrez, S.P. Chockalingam, C.J. Arguello, L. Palova, D. Nordlund, M.S. Hybertsen, D.R. Reichman, T.F. Heinz, P. Kim, A. Pinczuk, G.W. Flynn, A.N. Pasupathy, Science **333**, 999 (2011)
38. R. Lv, Q. Li, A.R. Botello-Mendez, T. Hayashi, B. Wang, A. Berkdemir, Q. Hao, A.L. Elas, R. Cruz-Silva, H.R. Gutierrez, Y.A. Kim, H. Muramatsu, J. Zhu, M. Endo, H. Terrones, J.-C. Charlier, M. Pan, M. Terrones, Sci. Rep. **2**, 586 (2012)
39. R. Balog, B. Jørgensen, L. Nilsson, M. Andersen, E. Rienks, M. Bianchi, M. Fanetti, E. Lægsgaard, A. Baraldi, S. Lizzit, Z. Sljivancanin, F. Besenbacher, B. Hammer, T.G. Pedersen, P. Hofmann, L. Hornekær, Nat. Mater. **9**, 315 (2010)
40. S. Roche, D. Mayou, Phys. Rev. Lett. **79**, 2518 (1997)
41. S. Roche, Phys. Rev. B **59**, 2284 (1999)
42. F. Ortmann, A. Cresti, G. Montambaux, S. Roche, Europhys. Lett. **94**, 47006 (2011)
43. S. Roche, N. Leconte, F. Ortmann, A. Lherbier, D. Soriano, J.-C. Charlier, Solid State Commun. **152**, 1404–1410 (2012)
44. A. Cresti, G. Grosso, G.P. Parravicini, Phys. Rev. B **76**, 205433 (2007)
45. A. Cresti, N. Nemec, B. Biel, G. Niebler, F. Triozon, G. Cuniberti, S. Roche, Nano Res. **1**, 361 (2008)
46. A. Lherbier, S.M.M. Dubois, X. Declerck, Y.M. Niquet, S. Roche, J.C. Charlier, Phys. Rev. B **86**, 075402 (2012)
47. N. Leconte, D. Soriano, S. Roche, P. Ordejon, J.C. Charlier, J.J. Palacios, ACS Nano **5**, 3987–3992 (2011)
48. M.A.H. Vozmediano, F. Guinea, Phys. Scr. **T146**, 014015 (2012)
49. D.C. Elias, R.V. Gorbachev, A.S. Mayorov, S.V. Morozov, A.A. Zhukov, P. Blake, L.A. Ponomarenko, I.V. Grigorieva, K.S. Novoselov, F. Guinea, A.K. Geim, Nat. Phys. **7**, 701 (2011)
50. Y. Barlas, K. Yang, A. MacDonald, Nanotechnology **23**, 052001 (2012)
51. K.S. Novoselov, V.I. Falko, L. Colombo, P.R. Gellert, M.G. Schwab, K. Kim, Nature **192**, 490 (2012)
52. F. Bonaccorso, Z. Sun, T. Hasan, A.C. Ferrari, Nat. Photonics **4**, 611 (2010)
53. F. Schwierz, Nat. Nanotechnol. **5**, 487 (2010)
54. Q. Bao, K.P. Loh, ACS Nano **6**, 3677 (2012)
55. T. Low, P. Avouris, ACS Nano **8**, 1086 (2014)
56. P. Avouris, Nano Lett. **10**, 4285 (2010)
57. X. Huang, Z. Zeng, Z. Fan, J. Liu, H. Zhang, Adv. Mater. **24**, 5979 (2012)
58. C. Lee, X. Wei, J.W. Kysar, J. Hone, Science **321**, 385 (2008)
59. K.S. Novoselov, A.K. Geim, S.V. Morozov, D. Jiang, Y. Zhang, S.V. Dubonos, I.V. Grigorieva, A.A. Firsov, Science **306**, 666 (2004)
60. A.A. Balandin, S. Ghosh, W. Bao, I. Calizo, D. Teweldebrhan, F. Miao, C.N. Lau, Nano Lett. **8**, 902 (2008)
61. R.R. Nair, P. Blake, A.N. Grigorenko, K.S. Novoselov, T.J. Booth, T. Stauber, N.M.R. Peres, A.K. Geim, Science **320**, 1308 (2008)
62. A. Pospischil, M. Humer, M.M. Furchi, D. Bachmann, R. Guider, T. Fromherz, T. Mueller, Nat. Photonics **7**, 892 (2013)
63. S. Stankovich, D.A. Dikin, G.H.B. Dommett, K.M. Kohlhaas, E.J. Zimney, E.A. Stach, R.D. Piner, S.T. Nguyen, R.S. Ruoff, Nature **442**, 282 (2006)
64. S. Bae, H. Kim, Y. Lee, X.F. Xu, J.S. Park, Y. Zheng, J. Balakrishnan, T. Lei, H.R. Kim, Y.I. Song, Y.J. Kim, K.S. Kim, B. Ozyilmaz, J.H. Ahn, B.H. Hong, S. Iijima, Nat. Nanotechnol. **5**, 574–578 (2010)
65. J. Ryu, Y. Kim, D. Won, N. Kim, J.S. Park, E.-K. Lee, D. Cho, S.-P. Cho, S.J. Kim, G.H. Ryu, H.-A.-S. Shin, Z. Lee, B.H. Hong, S. Cho, ACS Nano **8**, 950 (2014)
66. J. Riikonen, W. Kim, C. Li, O. Svensk, S. Arpiainen, M. Kainlauri, H. Lipsanen, Carbon **62**, 43 (2013)

67. L. Gao, G.-X. Ni, Y. Liu, B. Liu, A.H. Castro, Neto, K.P. Loh, Nature **505**, 190 (2014)
68. J. Kang, D. Shin, S. Bae, B.H. Hong, Nanoscale **4**, 5527 (2012)
69. L.H. Hess, M. Jansen, V. Maybeck, M.V. Hauf, M. Seifert, M. Stutzmann, I.D. Sharp, A. Offenhausser, J.A. Garrido, Adv. Mater. **23**, 5045 (2011)
70. H. Park, J.A. Rowehl, K.K. Kim, V. Bulovic, J. Kong, Nanotechnology **21**, 505204 (2010)
71. D. Pesin, A.H. MacDonald, Nat. Mater. **11**, 409 (2012)
72. P.Y. Huang, C.S. Ruiz-Vargas, A.M. van der Zande, W.S. Whitney, M.P. Levendorf, J.W. Kevek, S. Garg, J.S. Alden, C.J. Hustedt, Y. Zhu, J. Park, P.L. McEuen, D.A. Muller, Nature **469**, 389 (2011)
73. K. Kim, Z. Lee, W. Regan, C. Kisielowski, M.F. Crommie, A. Zettl, ACS Nano **5**, 2142 (2011)
74. O.V. Yazyev, S.G. Louie, Nat. Mater. **9**, 806 (2010)
75. A. Salehi-Khojin, D. Estrada, K.Y. Lin, M.-H. Bae, F. Xiong, E. Pop, R.I. Masel, Adv. Mater. **24**, 53 (2012)
76. O.C. Compton, S.T. Nguyen, Small **6**, 711 (2010)
77. X. Huang, Z. Yin, S. Wu, X. Qi, Q. He, Q. Zhang, Q. Yan, F. Boey, H. Zhang, Small **7**, 1876 (2011)
78. S. Park, R.S. Ruoff, Nat. Nanotechnol. **4**, 217 (2009)
79. Y. Zhu, S. Murali, W. Cai, X. Li, J.W. Suk, J.R. Potts, R.S. Ruoff, Adv. Mater. **22**, 3906 (2010)
80. C.N.R. Rao, A.K. Sood, K.S. Subrahmanyam, A. Govindaraj, Angew. Chem. Int. Ed. Engl. **48**, 7752 (2009)
81. C.K. Chua, M. Pumera, Chem. Soc. Rev. **43**, 291 (2014)
82. L.P. Biro, P. Lambin, New J. Phys. **15**, 035024 (2013)
83. P.T. Araujo, M. Terrones, M.S. Dresselhaus, Mater. Today **15**, 98 (2012)
84. M. Batzill, Surf. Sci. Rep. **67**, 83 (2012)
85. F. Banhart, J. Kotakoski, A.V. Krasheninnikov, ACS Nano **5**, 26–41 (2011)
86. J. Kotakoski, A.V. Krasheninnikov, U. Kaiser, J.C. Meyer, Phys. Rev. Lett. **106**, 105505 (2011)
87. L. Tapaszto, P. Nemes-Incze, G. Dobrik, K.J. Yoo, Ch. Hwang, L.P. Biro, Appl. Phys. Lett. **100**, 053114 (2012)
88. Y. Liu, B.I. Yakobson, Nano Lett. **10**, 2178 (2010)
89. D.L. Duong et al, Nature **490**, 235–239 (2012)
90. W. Yang, G. Chen, Z. Shi, C.-C. Liu, L. Zhang, G. Xie, M. Cheng, D. Wang, R. Yang, D. Shi, K. Watanabe, T. Taniguchi, Y. Yao, Y. Zhang, G. Zhang, Nat. Mater. **12**, 792 (2013)
91. J. An, E. Voelkl, J.W. Suk, X. Li, C.W. Magnuson, L. Fu, P. Tiemeijer, M. Bischoff, B. Freitag, E. Popova, R.S. Ruoff, ACS Nano **5**, 2433 (2011)
92. C.S. Ruiz-Vargas, H.L. Zhuang, P.Y. Huang, A.M. van der Zande, S. Garg, P.L. McEuen, D.A. Muller, R.G. Hennig, J. Park. Nano Lett. **11**, 2259 (2011)
93. R. Grantab, V.B. Shenoy, R.S. Ruoff, Science **330**, 946 (2010)
94. J. Zhang, J. Zhao, J. Lu, ACS Nano **6**, 2704 (2012)
95. Y. Wei, J. Wu, H. Yin, X. Shi, R. Yang, M. Dresselhaus, Nat. Mater. **11**, 759 (2012)
96. H.I. Rasool, C. Ophus, W.S. Klug, A. Zettl, J.K. Gimzewski, Nat. Commun. **4**, 2811 (2013)
97. G.-H. Lee, R.C. Cooper, S.J. An, S. Lee, A. van der Zande, N. Petrone, A.G. Hammerberg, C. Lee, B. Crawford, W. Oliver, J.W. Kysar, J. Hone, Science **340**, 1073 (2013)
98. K. Kim, V.I. Artyukhov, W. Regan, Y. Liu, M.F. Crommie, B.I. Yakobson, A. Zettl, Nano Lett. **12**, 293 (2012)
99. Z. Song, V.I. Artyukhov, B.I. Yakobson, Z. Xu, Nano Lett. **13**, 1829 (2013)
100. A.W. Tsen et al., Science **336**, 1143–1146 (2012)
101. L.A. Jauregui, H. Cao, W. Wu, Q. Yu, Y.P. Chen, Solid State Commun. **151**, 1100 (2011)
102. Q. Yu et al., Nat. Mater. **10**, 443–449 (2011)
103. Y. Hao, M.S. Bharathi, L. Wang, Y. Liu, H. Chen, S. Nie, X. Wang, H. Chou, C. Tan, B. Fallahazad, H. Ramanarayan, C.W. Magnuson, E. Tutuc, B.I. Yakobson, K.F. McCarty, Y.-W. Zhang, P. Kim, J. Hone, L. Colombo, R.S. Ruoff, Science **342**, 720 (2013)
104. Z. Yan, J. Lin, Z. Peng, Z. Sun, Y. Zhu, L. Li, C. Xiang, E.L. Samuel, C. Kittrell, J.M. Tour, ACS Nano **6**, 9110 (2012)

105. H. Zhou, W.J. Yu, L. Liu, R. Cheng, Y. Chen, X. Huang, Y. Liu, Y. Wang, Y. Huang, X. Duan, Nat. Commun. **4**, 2096 (2013)
106. L. Gan, Z. Luo, ACS Nano **7**, 9480 (2013)
107. L. Gao, W. Ren, H. Xu, L. Jin, Z. Wang, T. Ma, L.-P. Ma, Z. Zhang, Q. Fu, L.-M. Peng, X. Bao, H.-M. Cheng, Nat. Commun. **3**, 699 (2012)
108. G.H. Han, F. Gunes ans J. J. Bae, E. S. Kim, S. J. Chae, H.-J. Shin, J.-Y. Choi, D. Pribat, Y. H. Lee. Nano Lett. **11**, 4144 (2011)
109. Y.A. Wu, Y. Fan, S. Speller, G.L. Creeth, J.T. Sadowski, K. He, A.W. Robertson, C.S. Allen, J.H. Warner, ACS Nano **6**, 5010 (2012)
110. D. Geng, B. Luo, J. Xu, Y. Guo, B. Wu, W. Hu, Y. Liu, GYu. Adv, Funct. Mater. **24**, 1664 (2014)
111. D. Geng, B. Wu, Y. Guo, L. Huang, Y. Xue, J. Chen, G. Yu, L. Jiang, W. Hu, Y. Liu, Proc. Natl. Acad. Sci. USA **109**, 7992 (2012)
112. B. Hu, H. Ago, Y. Ito, K. Kawahara, M. Tsuji, E. Magome, K. Sumitani, N. Mizuta, K. Ikeda, S. Mizuno, Carbon N. Y. **50**, 57 (2012)
113. J.-S. Yu, D.-H. Ha, J.-H. Kim, Nanotechnology **23**, 395704 (2012)
114. D.W. Kim, Y.H. Kim, H.S. Jeong, H.-T. Jung, Nat. Nanotechnol. **7**, 29 (2012)
115. J.-H. Son, S.-J. Baeck, M.-H. Park, J.-B. Lee, C.-W. Yang, J.-K. Song, W.-C. Zin, J.-H. Ahn, Nat. Commun. **5**, 3484 (2014)
116. P. Nemes-Incze et al., Appl. Phys. Lett. **94**, 023104 (2011)
117. S. Lai, S. Kyu, Jang, Y. Jae Song, S. Lee. Appl. Phys. Lett. **104**, 043101 (2014)
118. J. Kotakoski, J.C. Meyer, Phys. Rev. B **85**, 195447 (2012)
119. D.W. Brenner, O. Shenderova, J. Harrison, S. Stuart, B. Ni, S. Sinnott, J. Phys.: Condens. Matter **14**, 783 (2002)
120. H. Berendsen, P.M. Postma, W. Gunsteren, A. DiNola, J. Haak, J. Chem. Phys. **81**, 3684 (1984)
121. S. Kurasch, J. Kotakoski, O. Lehtinen, V. Skakalova, J. Smet, C.E. Krill, A.V. Krasheninnikov, U. Kaiser, Nano Lett. **12**, 3168 (2012)
122. J. Martin, N. Akerman, G. Ulbright, T. Lohmann, J.H. Smet, K. von Klitzing, A. Yacoby, Nat. Phys. **4**, 144 (2008)
123. E.H. Hwang, S. Adam, S.D. Sarma, Phys. Rev. Lett. **98**, 18 (2007)
124. A.W. Robertson, A. Bachmatiuk, Y.A. Wu, F. Schaffel, B. Rellinghaus, B. Buchner, M.H. Rummeli, J.H. Warner, ACS Nano **5**, 6610 (2011)
125. P. Simonis, C. Goffaux, P. Thiry, L. Biro, P. Lambin, V. Meunier, Surf. Sci. **511**, 319 (2002)
126. J. Cervenka, C.F.J. Flipse, Phys. Rev. B **79**, 195429 (2009)
127. Z. Fei, A.S. Rodin, W. Gannett, S. Dai, W. Regan, M. Wagner, M.K. Liu, A.S. Mcleod, G. Dominguez, M. Thiemens, A.H.C. Neto, F. Keilmann, A. Zettl, R. Hillenbrand, M.M. Fogler, D.N. Basov, Nat. Nanotechnol. **8**, 821 (2013)
128. J.C. Koepke, J.D. Wood, D. Estrada, Z.-Y. Ong, K.T. He, E. Pop, J.W. Lyding, ACS Nano **7**, 75 (2013)
129. P. Nemes-Incze et al., Carbon N. Y. **64**, 178 (2013)
130. I. Vlassiouk, S. Smirnov, I. Ivanov, P.F. Fulvio, S. Dai, H. Meyer, M. Chi, D. Hensley, P. Datskos, N.V. Lavrik, Nanotechnology **22**, 275716 (2011)
131. J. Cervenka, M.I. Katsnelson, C.F.J. Flipse, Nat. Phys. **5**, 840 (2009)
132. X. Li, Y. Zhu, W. Cai, M. Borysiak, B. Han, D. Chen, R.D. Piner, L. Colombo, R.S. Ruoff, Nano Lett. **9**, 4359 (2009)
133. A. Turchanin, A. Beyer, C.T. Nottbohm, X. Zhang, R. Stosch, A. Sologubenko, J. Mayer, P. Hinze, T. Weimann, A. Golzhauser, Adv. Matter **21**, 1233 (2009)
134. Z.H. Ni, L.S. Ponomarenko, R.R. Nair, R. Yang, S. Anissimova, I.V. Grigorieva, F. Schedin, P. Blake, Z.X. Shen, E.H. Hill, K.S. Novoselov, A.K. Geim, Nano Lett. **10**, 3868 (2010)
135. J.-H. Chen, C. Jang, S. Xiao, M. Ishigami, M.S. Fuhrer, Nat. Nanotechnol. **3**, 206 (2008)
136. D.W. Boukhvalov, M.I. Katsnelson, Nano Lett. **8**, 4373–4379 (2008)
137. F. Ouyang, J. Xiao, H. Wang, H. Xu, J. Phys. Chem. C **112**, 12003 (2008)
138. N. Ghaderi, M. Peressi, J. Phys. Chem. C **114**, 21625 (2010)

139. L. Chen, J. Li, D. Li, M. Wei, X. Wang, Olid State Commun. **152**, 1985 (2012)
140. X. Qi, X. Guo, C. Zheng, Appl. Surf. Sci. **259**, 195 (2012)
141. S. Jandhyala, G. Mordi, B. Lee, G. Lee, C. Floresca, P.-R. Cha, J. Ahn, R.M. Wallace, Y.J. Chabal, M.J. Kim, L. Colombo, K. Cho, J. Kim, ACS Nano **6**, 2722 (2012)
142. G. Lee, B. Lee, J. Kim, K. Cho, J. Phys. Chem. C **113**, 14225 (2009)
143. A. Ferreira, X. Xu, C.-L. Tan, S. Bae, N.M.R. Peres, B.-H. Hong, B. Ozyilmaz, A.H. Castro Neto, Eur. Phys. Lett. **94**, 28003 (2011)
144. G.I. Mark, P. Vancso, C. Hwang, P. Lambin, L.P. Biro, Phys. Rev. B **85**, 125443 (2012)
145. O.V. Yazyev, F. Gargiulo, Nano Lett. **14**, 250 (2014)
146. P. Vancso, G.I. Mark, P. Lambin, A. Mayer, Y.-S. Kim, C. Hwang, L.P. Biro, Carbon **64**, 101 (2013)
147. E.F. Luis, F. Torres, S. Roche, J.C. Charlier, *Introduction to Graphene-Based Nanomaterials From Electronic Structure to Quantum Transport* (Cambridge, 2013)
148. K. Saloriutta, A. Uppstu, A. Harju, M.J. Puska, Phys. Rev. B **86**, 235417 (2012)
149. K.W. Clark et al., ACS Nano **7**, 7956–7966 (2013)
150. S. Roche, S.O. Valenzuela, J. Phys. D. Appl. Phys. **47**, 094011 (2014)
151. I. Meric, M.Y. Han, A.F. Young, B. Ozyilmaz, P. Kim, K. Shepard, Nat. Nanotechnol. **3**, 654–659 (2008)
152. D. Jimenez, Nanotechnology **19**, 345204 (2008)
153. D. Jimenez, IEEE Trans. Electron Devices **58**, 4377–4383 (2011)
154. T. Fang, A. Konar, H. Xing, D. Jena, Appl. Phys. Lett. **91**, 092109 (2007)
155. F. Schwierz, Proc. IEEE **101**, 1567–1584 (2013)
156. Y.-M. Lin, A. Valdes-Garcia, S.-J. Han, D.B. Farmer, I. Meric, Y. Sun, Y. Wu, C. Dimitrakopoulos, A. Grill, P. Avouris, K.A. Jenkins, Science **332**, 1294–1297 (2011)
157. H. Wang, A. Hsu, J. Wu, J. Kong, T. Palacios, IEEE Electron Device Lett. **31**, 906–908 (2010)
158. H. Wang, D. Nezich, J. Kong, T. Palacios, IEEE Electron Device Lett. **30**, 547–549 (2009)
159. X.S. Li, W.W. Cai, J.H. An, S. Kim, J. Nah, D.X. Yang, R. Piner, A. Velamakanni, I. Jung, E. Tutuc, S.K. Banerjee, L. Colombo, R.S. Ruoff, Science **324**, 1312–1314 (2009)
160. A. Reina, X. Jia, J. Ho, D. Nezich, H. Son, V. Bulovic, M.S. Dresselhaus, J. Kong, Nano Lett. **9**, 30 (2009)
161. X.S. Li, C.W. Magnuson, A. Venugopal, J.H. An, J.W. Suk, B.Y. Han, M. Borysiak, W.W. Cai, A. Velamakanni, Y.W. Zhu, L.F. Fu, E.M. Vogel, E. Voelkl, L. Colombo, R.S. Ruoff, Nano Lett. **10**, 4328–4334 (2010)
162. Y. Wu, Y.-M. Lin, A.A. Bol, K.A. Jenkins, F. Xia, D.B. Farmer, Y. Zhu, P. Avouris, Nature **472**, 74–78 (2011)
163. D.B. Farmer, A. Valdes-Garcia, C. Dimitrakopoulos, P. Avouris, Appl. Phys. Lett. **101**, 143503 (2012)
164. M.C. Lemme, T. Echtermeyer, M. Baus, B.N. Szafranek, M. Schmidt, H. Kurz, ECS Trans. **11**, 413 (2007)
165. D.S.L. Abergel, V. Apalkov, J. Berashevich, K. Ziegler, T. Chakraborty, Adv. Phys. **59**, 261 (2010)
166. V. Arkady, Krasheninnikov and Florian Banhart. Nat. Mater. **6**, 723 (2007)
167. J. Moser, H. Tao, S. Roche, F. Alsina, C.M. Sotomayor Torres, A. Bachtold, Phys. Rev. B **81**, 205445 (2010)
168. N. Leconte, J. Moser, P. Ordejon, H. Tao, A. Lherbier, A. Bachtold, F. Alzina, C.M. Sotomayor-Torres, J.-C. Charlier, S. Roche, ACS Nano **4**, 4033 (2010)
169. D. Teweldebrhan, A.A. Balandin, Appl. Phys. Lett. **94**, 013101 (2009)
170. H. Tao et al., J. Phys. Chem. C **115**, 18257 (2011)
171. Y.-B. Zhou et al., J. Chem. Phys. **133**, 234703 (2010)
172. V. Kapko, D.A. Drabold, M.F. Thorpe, Phys. Stat. Solidi B **247**, 1197–1200 (2010)
173. E. Holmström et al., Phys. Rev. B **84**, 205414 (2011)
174. Y. Li, F. Inam, F. Kumar, M.F. Thorpe, D.A. Drabold, Phys. Stat. Solidi B **248**, 2082–2086 (2011)
175. F. Wooten, K. Winer, D. Weaire, Phys. Rev. Lett. **54**, 1392 (1985)

176. H. He, Ph.D. thesis. Michigan State University, 1985
177. A.J. Stone, D.J. Wales, Chem. Phys. Lett. **128**, 501 (1986)
178. P.N. Keating, Phys. Rev. **145**, 637 (1966)
179. R. Kubo, Rep. Prog. Phys. **29**, 255 (1966)
180. M. Buttiker et al., Phys. Rev. B **31**, 6207 (1985)
181. K.S. Thygesen, K.W. Jacobsen, Phys. Rev. B **72**, 033401 (2005)
182. R. Rurali, F.D. Novaes, P. Ordejón, ACS Nano **4**, 7596 (2010)
183. P.A. Lee, T.V. Ramakrishnan, Rev. Mod. Phys. **57**, 2 (1985)

Chapter 5
Spin Transport in Disordered Graphene

Carbon has a weak atomic SOC, so graphene is expected to have long spin relaxation time and phase coherence lengths. However, the spin injection measurements based on a non-local spin valve geometry [1–3] revealed surprisingly short spin relaxation times of only about 100–200 ps, being only weakly dependent on the charge density and temperature. The longest spin relaxation time has been measured up to now is also in the order of a few ns [4]. There are many explanations for short spin relaxation times in graphene. Some explanations are related to enhanced SOC induced by adatoms in graphene sheet or by the breaking inversion symmetry due to the electric field created by substrate. Another possibility could be the Gauge field due to ripples [5] which induces an effective magnetic field $B_\perp$ perpendicular to the graphene sheet. There is also another explanation saying that the formation of sp^3 hybridization enhances local SOC [6] which leads to fast spin relaxation. However, these theoretical results couldn't give satisfying explanations for experimental data to date.

In this section, we perform some theoretical calculations to investigate the spin relaxation in ultra clean graphene, and we propose a new mechanism for spin relaxation in graphene which is related to the disconnection of spin and momentum close to the Dirac point. At the end of this Chapter, some results of the effect of the segregation of strong-SOC adatoms on graphene on the QHE are shown.

5.1 Spin Transport in Graphene: Pseudospin Driven Spin Relaxation Mechanism

5.1.1 Introduction

The electronic properties of monolayer graphene strongly differ from those of two-dimensional metals and semiconductors in part because of inherent electron-hole band structure symmetry and a particular density of states which vanishes at the Dirac

© Springer International Publishing Switzerland 2016
D.V. Tuan, *Charge and Spin Transport in Disordered Graphene-Based Materials*, Springer Theses, DOI 10.1007/978-3-319-25571-2_5

point [6]. Additionally, the sublattice degeneracy and honeycomb symmetry lead to eigenstates that hold an additional quantum (Berry's) phase, associated with the so-called pseudospin quantum degree of freedom. All of these electronic features are manifested through the Klein tunneling phenomenon [7], WAL [8] or the anomalous QHE [9]. The possibility of using the pseudospin as a means to transport and store information has also been theoretically proposed [10, 11]. There, the role of the pseudospin is equivalent to that of the spin in spintronics, such as in the pseudospin analogue of the giant magnetoresistance in bilayer graphene [11].

Even though pseudospin-related effects drive most of the unique transport signatures of graphene, the role of the pseudospin on the spin relaxation mechanism has not been explicitly addressed and quantified. Pseudospin and spin dynamics are usually perceived as decoupled from each other, with pseudospin lifetimes being much shorter and pseudospin dynamics much faster than those for spins. However, this picture breaks down in the vicinity of the Dirac point, a region that is usually out of reach of perturbative approaches and that is particularly relevant for experiments, because Fermi energies can only be shifted by about 0.3 eV via electrostatic gating. Moreover, in the presence of SOC, spin couples to orbital motion, and therefore to pseudospin [12], so that spin and pseudospin dynamics should not be treated independently.

The reason for overlooking the role of the pseudospin on the spin dynamics is perhaps rooted in the fact that the spin transport properties appear remarkably similar to those found in common metals and semiconductors [13]. Indeed, spin precession measurements in nonlocal devices result in experimental signatures that would be indistinguishable from those obtained in a metal such as aluminium [14], or a semiconductor such as GaAs [15], with extracted spin relaxation times τ_s that are also typically of the same order of magnitude (a few nanoseconds or lower). Spin relaxation in graphene has therefore been interpreted using the conventional experimental manifestations of either the EY or DP mechanism [16–20]. In the EY scenario, the spin relaxation time is determined by the spin mixing of carriers and the SOC of the scattering potential, and thus it is usually assumed to be proportional to the momentum relaxation time as $\tau_s \approx \alpha \cdot \tau_p$, with $\alpha \gg 1$ (for instance in alkali metals $\alpha \sim 10^4 - 10^6$) [13]. In contrast, in the DP mechanism spin precesses about an effective magnetic field whose orientation is fixed by the momentum direction during free propagation of electrons. Such orientation changes at each scattering event, which results in a different scaling behavior as $1/\tau_s^{DP} \sim \Omega^2 \tau_p$ [13] (with Ω the average magnitude of the intrinsic Larmor frequency over the momentum distribution). Experimental estimates of τ_s and τ_p are generally obtained in a phenomenological way by fitting the experimental resistivity curves to the theoretical formula obtained using semi-classical transport equations [1, 17]. However, this phenomenological analysis is not well connected with the microscopic interpretation. First of all, the weak SOC in graphene would suggest τ_s in the microsecond range [21, 22], in clear disagreement with experimental data. In addition, the τ_s estimated in high-mobility graphene with long mean free paths remains unsatisfactorily interpreted with a single relaxation mechanism, say EY or DP [18, 23, 24]. The suppression of τ_s in clean graphene has been tentatively associated to an enhanced (intrinsic or extrinsic) SOC due to mechanical deformations such as ripples, or unavoidable adatoms incorporated during the device fabrication process, but the ultimate and microscopic nature of spin relaxation at play remains controversial and elusive.

Here, we unravel a spin relaxation mechanism for nonmagnetic samples that follows from an entanglement of spin and pseudospin degrees of freedom driven by random SOC, which makes it unique to graphene and is markedly different from conventional processes. We show that the mixing between spin and pseudospin-related Berry's phases results in unexpectedly fast spin dephasing, even when approaching the ballistic limit, and leads to increasing spin relaxation times away from the Dirac point, as observed experimentally. This hitherto unknown phenomenon points towards revisiting the origin of the small spin relaxation times found in graphene, where SOC can be caused by adsorbed adatoms, ripples or even the substrate. It also opens new perspectives for spin manipulation using the pseudospin degree of freedom (or vice versa), a tantalizing quest for the emergence of radically new information storage and processing technologies.

5.1.2 Spin Relaxation in Gold-Decorated Graphene

In the following, we explore spin characteristics in graphene by investigating the effect of weak perturbation induced by low densities of ad-atoms (down to 10^{12} cm^{-2}), which introduce a random Rashba field in real space but vanishingly small intervalley scattering, yielding long mean free paths. Here, for typical electron densities within $[10^{10}, 10^{12}]$ cm^{-2}, the Fermi wavelength ($\lambda_F = 2\sqrt{\pi/n}$, n the charge density) lies between 20 and 200 nm and thus exceeds the mean separation between adatoms ($\sim$10 nm) where spin-orbit scattering occurs, therefore questioning the use of a standard semiclassical description. To study spin dynamics (and spin relaxation), we use a non-perturbative method by solving the full time-dependent evolution of initially spin polarized wavepackets, either through a direct diagonalization of a continuum model, or a real space algorithm for a microscopic disorder model, defined in a TB basis. We describe the system of a graphene monolayer functionalized with a random distribution of adatoms. The electronic structure of clean graphene is captured by the usual π-π* orthogonal TB model (with a single p_z-orbital per carbon site, zero onsite energies and nearest neighbors hopping γ_0). The presence of non-magnetic adatoms randomly adsorbed at the hollow positions on the graphene sheet introduces additional local SOC terms (Fig. 5.1a, b), defined as [25].

$$\mathcal{H} = -\gamma_0 \sum_{\langle ij \rangle} c_i^+ c_j + \frac{2i}{\sqrt{3}} V_I \sum_{\langle\langle ij \rangle\rangle \in \mathcal{R}} c_i^+ \vec{s} \cdot (\vec{d}_{kj} \times \vec{d}_{ik}) c_j$$
$$+ i V_R \sum_{\langle ij \rangle \in \mathcal{R}} c_i^+ \vec{z} \cdot (\vec{s} \times \vec{d}_{ij}) c_j - \mu \sum_{i \in \mathcal{R}} c_i^+ c_i \qquad (5.1)$$

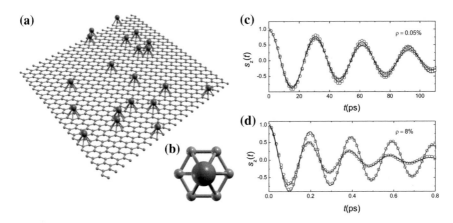

Fig. 5.1 Spin Dynamics in disordered graphene. **a** Ball-and-stick model of a random distribution of adatoms on *top* of a graphene sample. **b** *Top view* of the gold adatom sitting on the center of an hexagon. **c, d** Time-dependent projected spin polarization $S_z(E, t)$ of charge carriers (symbols) initially prepared in an out-of-plane polarization (at Dirac point (*red curves*) and at $E = 150$ meV (*blue curves*)). Analytical fits are given as *solid lines* (see text). Parameters are $V_I = 0.007\gamma_0$, $V_R = 0.0165\gamma_0$, $\mu = 0.1\gamma_0$, $\rho = 0.05\%$ (**c**) and $\rho = 8\%$ (**d**)

The first term is the nearest neighbor hopping term with $\gamma_0 = 2.7$ eV. The second term is a complex next nearest neighbor hopping term which represents the intrinsic SOC induced by the adatoms, with $\vec{d}_{kj}$ and $\vec{d}_{ik}$ the unit vectors along the two bonds connecting second neighbors, $\vec{s}$ is a vector defined by the Pauli matrices (s_x, s_y, s_z), and V_I the intrinsic SOC strength. The third term describes the Rashba SOC which explicitly violates $\vec{z} \rightarrow -\vec{z}$ symmetry, with $\vec{z}$ being a unit vector normal to the graphene plane and V_R the Rashba SOC parameter. The last term is the potential shift μ associated with the carbon atoms in the random plaquettes $\mathcal{R}$ adjacent to adatoms (Fig. 5.1b). Such shift is due to weak electrostatic effects that arise from charge redistribution induced very locally around the adatom [25].

A Rashba splitting has been observed experimentally at the graphene/nickel and graphene/gold (Au) interfaces with spin splitting of up to 100 meV [26, 27]. Gold and nickel as well as other materials like titanium, cobalt or chromium, are usually present during the fabrication of the nonlocal spin valves that are used to determine τ_s and likely leave residues on the exposed graphene surface. Hereafter, we consider the case of Au adatoms whose influence on the transport properties of graphene has been studied experimentally [28]. The TB parameters to describe both intrinsic and Rashba SOCs induced by such adatoms are extracted from *ab-initio* calculations [27]. Based on such parameters, we explore how the spin relaxation times scale as a function of the adatom density and adatom-induced local potential shift.

The spin dynamics in graphene are investigated by computing the time-dependence of the spin polarization defined by (See Sect. 3.2.2 for technical details)

$$\vec{S}(E, t) = \frac{\langle \Psi(t) | \vec{s}\delta(E - \mathcal{H}) + \delta(E - \mathcal{H})\vec{s} \, | \Psi(t) \rangle}{2\langle \Psi(t) | \delta(E - \mathcal{H}) | \Psi(t) \rangle} \tag{5.2}$$

and assuming that spins are initially injected out-of-plane (z direction), i.e. $|\Psi(t = 0)\rangle = |\psi_\uparrow\rangle$. The time evolution of the wavepackets $|\Psi(t)\rangle$ is obtained by solving the time-dependent Schrödinger equation. We focus on the expectation value of the spin z-component $S_z(E, t)$. Figure 5.1 shows the typical behavior of $S_z(E, t)$ for two selected energies (at the Dirac point and at $E = 150$ meV) and two adatom densities $\rho = 0.05$ % (about 10^{12} adatoms per cm^2) (c) and $\rho = 8$ % (d). The time dependence of $S_z(E, t)$ is very well described by $\cos(2\pi t/T_\Omega)e^{-t/\tau_s}$, introducing the spin precession period T_Ω and the spin relaxation time τ_s, which are extracted from fitting the numerical simulations (solid lines). The time dependence of the modulus of the full spin polarization vector $|\vec{S}| = |(\langle s_x\rangle, \langle s_y\rangle, \langle s_z\rangle)|$ also exhibits an unambiguous signature of spin relaxation (See Sect. 5.1.3). Figure 5.2 gives τ_s and T_Ω extracted from the fits of $S_z(E, t)$ for varying adatom density. One first observes that the spin precession period is energy independent and is precisely equal to $T_\Omega = \pi\hbar/\bar{\lambda}_R$ (with $\bar{\lambda}_R = 3\rho V_R$ an average SOC strength) even for the lowest coverage, which

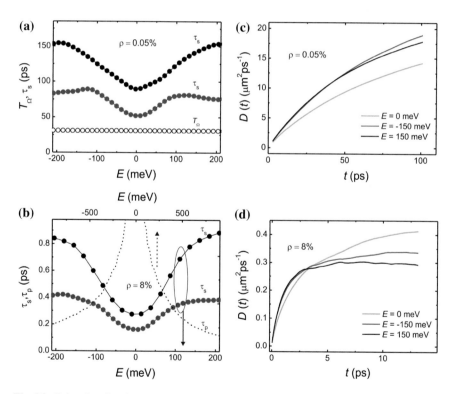

Fig. 5.2 Spin relaxation times and transport mechanisms. Spin relaxation times (τ_s) for $\rho = 0.05$ % (a) and $\rho = 8$ % (b). *Black (red)* solid symbols indicate τ_s for $\mu = 0.1\gamma_0$ ($\mu = 0.2\gamma_0$). T_Ω versus E is also shown (*open symbols*). τ_p (*dotted line* in (b)) is shown over a wider energy range (top x-axis) in order to stress the divergence around $E = 0$ ($\mu = 0.2\gamma_0$). We cannot evaluate τ_p below 100 meV, since the diffusive regime is not established within our computational reach. Panels (c) and (d): Time dependent diffusion coefficient $D(t)$ for $\rho = 0.05$ % and $\rho = 8$ % with $\mu = 0.2\gamma_0$

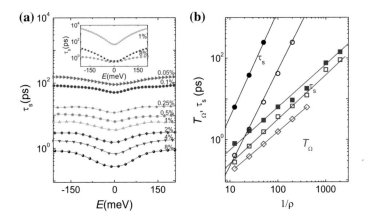

Fig. 5.3 Spin relaxation times deduced from the continuum and microscopic models. **a** Spin relaxation times (τ_s) for varying ρ between 0.05 and 8 % extracted from the microscopic model (with $\mu = 0.1\gamma_0$). *Inset* τ_s values using the continuum model for $\rho = 1$ and 8 % (*filled symbols*). A comparison with the microscopic model (with $\mu = 0$) is also given for $\rho = 8\%$ (*open circles*). **b** Scaling behavior of T_Ω and τ_s versus $1/\rho$. The T_Ω values obtained with the microscopic (resp. continuum) model are given by red diamonds (resp. *red solid lines*). τ_s values for the microscopic model (*blue squares*) and the continuum model (*black circles*) are shown for two selected energies $E = 150$ meV (*solid symbols*) and $E = 0$ (*open symbols*). *Solid lines* are here guides to the eye

agrees with the estimate based on the continuum model [21] (See Eq. (2.52)). In contrast, the spin relaxation time displays a significant energy dependence. A V-shape is obtained for low energy, with τ_s being minimal at the Dirac point with values ranging from 0.1 to 200 ps when tuning the adatom density from 8 to 0.05 % (as given in Fig. 5.3a, main frame). Based on the observed scaling $\tau_s \sim 1/\rho$ (see Fig. 5.3b), one can further extrapolate the spin relaxation times for even smaller defect density, obtaining $\tau_s \sim 1 - 10$ ns for adsorbate densities decreasing from 10^{11} cm^{-2} down to 10^{10} cm^{-2}. The obtained V-shaped energy dependence and the absolute values of τ_s are remarkably similar to those reported experimentally [1, 16, 17, 28].

The faster relaxation at the Dirac point is actually evident in Fig. 5.1c and d. The reason for this behaviour is the decrease of the coupling between the pseudospin and momentum and the increasing dominance of the SOC interaction, which leads to spin-pseudospin entanglement. The details of the entanglement are further described in Eq. (5.3) below and in the Sect. 5.1.3

As discussed above, the usual approach to discriminate between conventional EY and DP relaxation mechanisms in metals and semiconductors is to scrutinize the scaling of τ_s versus τ_p. Such procedure does not necessarily apply if the dominant processes that lead to momentum and scattering relaxation are not the same. For instance, in monolayer transition-metal dichalcogenides, it was demonstrated that the carrier scattering by flexural phonons leads to fast spin flips but not to momentum scattering and, therefore, the spin transport is decoupled from the carrier mobility. In the following discussion, we show that simple EY or DP scaling is also not suitable to describe our findings.

Within our microscopic calculations, we analyze the time-dependence of the diffusion coefficient for varying energies and ad-atom densities (Fig. 5.2c, d). For the lowest impurity density (0.05 %, Fig. 5.2c), regardless of the considered energy, $D(E, t)$ is seen to increase in time with no sign of saturation within our computational capability, indicating a ballistic-like regime for the considered timescales. Only for the largest ad-atom density (8 %) does $D(t)$ eventually saturate at high enough energies (above 100 meV, $D(t) \rightarrow D_{max}$), allowing for the evaluation of the transport time using $\tau_p(E) = D_{max}(E)/2v^2(E)$ (see dashed lines in Fig. 5.2b). A sharp increase of τ_p is seen when approaching the Dirac point, where τ_s reaches its minimum value, with $\tau_s \ll \tau_p$. This energy dependence in τ_p is not unique to gold ad-atoms but has also been observed for other types of disorder with a weak intervalley scattering contribution, such as epoxide defects or long range scatterers [29]. As seen in Fig. 5.3b, $\tau_s \sim 1/\rho$, which does not allow us to discriminate between EY and DP processes. However, the absolute values of τ_s and τ_p (with $\tau_s \ll \tau_p$) are a clear manifestation of the breakdown of the typical scaling associated to both mechanisms. Even the unconventional DP regime described in Ref. [13] for the case of $\tau_p/T_\Omega \geq 1$ where $1/\tau_s \sim \Delta\Omega$ (with $\Delta\Omega$ an effective width of the distribution of precession frequencies) cannot account for the observation that a weak variation in the local disorder affects the absolute values of τ_s (while ρ is unchanged) as observed in Fig. 5.2. Here local disorder is monitored by the μ parameter. (Although μ belongs to the TB parameterization of the adatom, we use it temporarily to increase local disorder.) In fact, its value could slightly change when modifying the substrate screening or in presence of a more strongly bonded adsorbant than Au. As a consequence of the above findings, the spin relaxation mechanism at play is incompatible with both the EY and the DP mechanisms, a fact which could shed new light on the current debate on the microscopic nature of spin relaxation in clean graphene [18, 23, 24].

We now further study the origin of the τ_s minimum at low energy and its unconventional scaling with τ_p. Given that our simulations with the microscopic model give $\tau_s \ll \tau_p$, we further explore the low-energy spin dynamics with an effective continuum model, in which the spin-orbit scattering is treated as a homogeneous potential [21]. We solve the Dirac equation in the continuum model by using a 4×4 effective Hamiltonian taking into account the pseudospin degree of freedom

$$h(\vec{k}) = h_0(\vec{k}) + h_R(\vec{k}) + h_I(\vec{k}) \tag{5.3}$$

While the hopping from three nearest neighbors $h_0(\vec{k}) = \hbar v_F(\zeta \sigma_x k_x + \sigma_y k_y) \otimes 1_s$ dominates at high energy and vanishes at the Dirac point ($\zeta = \pm 1$ for K and K' valleys, $\vec{\sigma}$ are pseudospin Pauli matrices and 1_s is a 2×2 identity matrix), the intrinsic SOC $h_I(\vec{k}) = \bar{\lambda}_I \zeta [\sigma_z \otimes s_z]$ and the Rashba interaction $h_R(\vec{k}) = \bar{\lambda}_R (\zeta [\sigma_x \otimes s_y] - [\sigma_y \otimes s_x])$ play an extremely important role at the Dirac point, where the coupling between spin and pseudospin becomes predominant, and governs the quantum dynamics and dephasing of the wavepackets as described below.

Within the continuum model spin relaxation is achieved by introducing an *ad-hoc* energy broadening. We use an initially z-polarized state for injection and consider only the K valley. A certain density of Au impurities (inducing local SOCs) is described by the effective SOCs $\bar{\lambda}_R = 3\rho V_R$ and $\bar{\lambda}_I = 3\sqrt{3}\rho V_I$. Note that no additional local (static) scattering potential is introduced here ($\mu = 0$). By computing the spin dynamics of initially spin-polarized wavepackets, one also obtains a spin relaxation effect defined by the two timescales T_Ω and τ_s (See Sect. 5.1.3).

It is instructive to contrast the results of the continuum model (Fig. 5.3a, inset) with those from the microscopic model (Fig. 5.3a, main frame). Although the spin precession period T_Ω obtained by both models is identical (Fig. 5.3b) and the energy dependence of τ_s is similar, the absolute values of τ_s differ substantially, especially in the high energy regime, where τ_s is clearly overestimated using the continuum model. Of key importance, such difference becomes increasingly large upon decreasing the ad-atom density because τ_s presents a different scaling with defect density (see Fig. 5.3b). This clearly evidences the importance of disorder, as introduced by the random distribution of impurities, and illustrates the limits of a phenomenological approach using the continuum model for quantitative comparison with experimental data. Notwithstanding, the qualitative agreement between both models (particularly for high coverage) and the weak momentum relaxation effects observed in the microscopic model (as seen in the long τ_p) suggest some generality in the unconventional spin relaxation observed near the Dirac point.

To further substantiate the origin of the spin relaxation, we scrutinize the spin and pseudospin dynamics of wavepackets using the continuum model. Pseudospin is intrinsically related to the graphene sublattice degeneracy and, as long as valley mixing is negligible, pseudospin is aligned in the direction of the momentum at high energy ($h_0(\vec{k})$ dominates the Hamiltonian (5.3)). The Rashba spin-orbit term $h_R(\vec{k})$ entangles spin $\vec{s}$ with the lattice pseudospin $\vec{\sigma}$, overriding the locking rule between pseudospin and momentum since $h_0(\vec{k})$ becomes vanishingly small in the vicinity of the Dirac point (see Sect. 5.1.3) [12, 19].

Figure 5.4 highlights the spin dynamics at different chosen energies $E = 0$, $E = -5\,\text{meV}$ (low energy) and $E = 130\,\text{meV}$ (high energy), which are representative of the underlying physics (note that no relaxation takes place for fixed energy, thus the requirement of the *ad-hoc* broadening). At high energy, the spin precesses quite regularly as seen in Fig. 5.4a, which shows an oscillatory pattern of $S_z(t)$ dominated by a single period $T_\Omega = \pi\hbar/\bar{\lambda}_R = 0.19$ ps. The spin precession occurs about an effective magnetic field B_R dictated by the Rashba interaction and pointing tangentially to the Fermi circle (as seen from the precession from blue to pink in right panels from t_1 to t_4). In contrast, the pseudospin $\langle\vec{\sigma}(t)\rangle$ points approximately in the same direction of the momentum (evolving from green to orange). Its oscillatory pattern is driven by the Rashba period T_Ω together with a superimposed and more rapid oscillation (described in the Sect. 5.1.3).

The situation at low energy (Fig. 5.4b, c) is markedly different. We observe a highly unconventional spin and pseudospin motion which is analyzed more closely for the spin and pseudospin z-components at two low energies (at the Dirac point and at $E = -5$ meV). In contrast to the high-energy case, the amplitude of the pseudospin oscillation is strongly enhanced since pseudospin is no longer locked with momentum but starts to precess about an effective pseudo-magnetic field. The

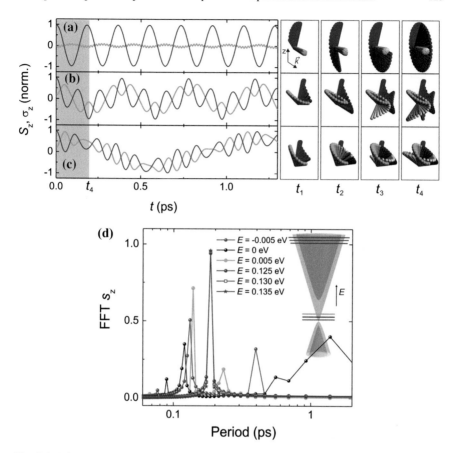

Fig. 5.4 Spin and pseudospin dynamics in graphene with $\rho = 8\,\%$ of adatoms **a–c**. Time dependence of spin-polarization S_z (*blue*) and pseudospin polarization σ_z (*green*) in z projection for energies $E = 130\,\text{meV}$ (**a**), $E = 0$ (**b**), and $E = -5\,\text{meV}$ (**c**). Note that all quantities are normalized to their maximum value to better contrast them in the same scale. *Right panels* show the time evolution for both spin (from *blue* to *pink*) and pseudospin (from *green* to *orange*). The snapshots are taken at different times from t_1 to t_4 sampling the shaded regions in (**a**)–(**c**). **d** Fourier transform of $S_z(t)$ plotted over oscillation period, and showing non-dispersive spectra at high energy (between $E = 125, 130$ and $135\,\text{meV}$). Low-energy spectra (for $E = -5, 0$ and $5\,\text{meV}$) change strongly with energy (dispersive) showing a gradual reduction and blue shift of the original Rashba peak at about $0.19\,\text{ps}$ and the appearance of additional features

pseudo-magnetic field depends strongly on the spin orientation, thus yielding complex time-dependent dynamics of spin and pseudospin (see right panels of Fig. 5.4 corresponding to 5.4b, c). Such an effect derives from the increased pseudospin precession period $T_0^{\text{ps}} = \pi\hbar/E$ (about B_0^{ps}), which decreases significantly at low energy. Therefore $\langle\sigma_i\rangle$ can no longer be replaced by its time average $\overline{\langle\sigma_i\rangle}$ (in contrast to the high-energy situation, see Sect. 5.1.3), which in consequence also holds for the Rashba field B_R. The time dependence of B_R with variability on a timescale similar to the Rashba period leads then to strong non-linear dynamics of spin and pseudospin motion. As a result of such coupled dynamics, the spin precession cannot be described by a single period T_Ω as becomes evident from the complex Fourier spectra of $S_z(t)$ in Fig. 5.4d. The time dependence of B_R includes also changes of its direction, thus impacting the pseudospin and lifting the pseudospin-momentum locking. Both of these effects finally produce a joint spin/pseudospin motion prohibiting the de-coupling of driving forces (B_0^{ps}, B_R) that was possible at higher energies.

While the continuum model provides qualitative insight into the spin-pseudospin coupling and entanglement of their corresponding wavefunctions, the microscopic model enables the quantification of spin relaxation times for a given microscopic disorder. By scrutinizing the general form of the spin polarization (Eq. (5.2)), a simple understanding of the spin relaxation mechanism can be drawn. In the microscopic model, the propagation of an initially spin-polarized wavepacket $|\psi_\uparrow(t = 0)\rangle$, is driven by the evolution operator $e^{-i\mathcal{H}t/\hbar}|\psi_\uparrow(t = 0)\rangle$, with $\mathcal{H}$ consisting of the clean graphene term plus the SOC potential, which acts as a local (and random) perturbation on the electron spin. The time-dependence of the total spin polarization results from the accumulated dephasing along scattering trajectories developed under the evolution operator. As the distribution of scattering centers is random in space, all different trajectories accumulate different phase shifts in their wavefunctions (each being the result of local spin/pseudospin coupling and disorder potential). When phase shifts for up and down components average out, the spin polarization of $|\psi_\uparrow(t = 0)\rangle$ is lost.

5.1.3 Further Discussion

Low-energy effective Hamiltonian and analysis of electronic states close to the Dirac point

To illustrate that spin and pseudospin are fully entangled for certain states close to the Dirac point, we calculate the band structure and the modulus of the spin polarization vector $|\vec{S}(\vec{k})|$. Figure 5.5 shows the computed band structure obtained by diagonalizing the Kane-Mele-Rashba Hamiltonian (Eq. (5.3)) for 8 % gold adatom coverage. The Rashba term induces a counter-propagating spin texture in the k_x, k_y plane that tends to vanish close to the Dirac point as [12]:

$$\vec{S}_{\nu\mu}(\vec{k}) = \frac{\mu\hbar v_F(\vec{k} \times \vec{z})}{\sqrt{\lambda_R^2 + \hbar^2 v_F^2 k^2}} \tag{5.4}$$

We further calculate the modulus of the spin polarization vector $|\vec{S}| = |(\langle s_x \rangle, \langle s_y \rangle, \langle s_z \rangle)|$ from the eigenstates of the full Hamiltonian in Eq. (5.3) with both intrinsic and Rashba SOC

$$\Psi_{\vec{k},\pm} = \left[\begin{pmatrix} c_{A,\uparrow} \\ c_{B,\uparrow} \end{pmatrix} \otimes | \uparrow \rangle \pm i \begin{pmatrix} c_{A,\downarrow} \\ c_{B,\downarrow} \end{pmatrix} \otimes | \downarrow \rangle \right] e^{i\vec{k}\vec{r}}. \tag{5.5}$$

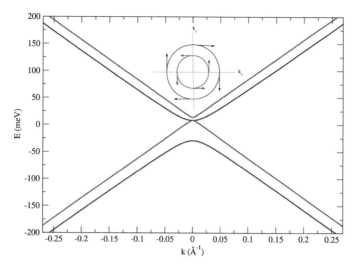

Fig. 5.5 Band structure calculated using the Kane-Mele-Rashba model for 8 % adatom concentration. The *inset* shows the typical Rashba-like spin texture for the conduction bands

In presence of the Rashba SOC term, Bloch states with well-defined spin polarization are no longer eigenstates of the complete Hamiltonian [19]. The clear signature of spin-pseudospin entanglement is found at low energies ($\vec{k} \to 0$), for which we get the following solutions

$$\Psi_{\vec{k},\pm}^{I} = \begin{pmatrix} 0 \\ 1 \end{pmatrix} \otimes | \uparrow \rangle \pm \begin{pmatrix} i \\ 0 \end{pmatrix} \otimes | \downarrow \rangle \tag{5.6}$$

$$\Psi_{\vec{k},\pm}^{II} = \begin{pmatrix} 1 \\ 0 \end{pmatrix} \otimes | \uparrow \rangle \pm \begin{pmatrix} 0 \\ i \end{pmatrix} \otimes | \downarrow \rangle. \tag{5.7}$$

In both cases, a change in sublattice (pseudospin) index entails a change in spin index. This means that at low energy spin and pseudospin are completely locked and $|\vec{S}| \approx 0$. The situation is different for high energies ($|\vec{k}| > 0$), when pseudospin-momentum coupling comes into play, all coefficients become equally weighted ($|c_{\sigma,s}| \approx 0.5$) and spin and pseudospin are unlocked leading to $|\vec{S}| \approx 1$.

Such energy dependence is shown in more details in Fig. 5.6, where the spin polarization $|\vec{S}|$ of the states in the two first conduction bands are computed by diagonalizing the effective Hamiltonian (Eq. (5.3)) for ad-atom concentrations $\rho = 25\%$ (1/4 ML gold coverage as reported by Marchenko et al. [27]) and $\rho = 8\%$ (which allows to make a connection with the microscopic model results in Fig. 5.7). The lower conduction-band states are completely entangled close to the Dirac point (red curves), but become disentangled at relatively low energies 25 and 100 meV for respectively low and high ad-atom densities (see vertical dashed lines). Interestingly, above these energies, the eigenstates of the second conduction band (blue curves) come into play with a stronger spin/pseudospin entanglement ($|\vec{S}| \ll 1$) even for high energy values: $E \sim 150$ meV for $\rho = 8\%$ and $E \approx 300$ meV for $\rho = 25\%$.

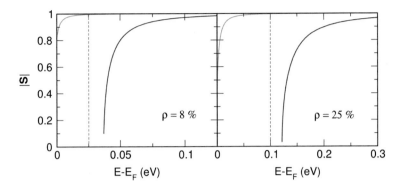

Fig. 5.6 Energy dependence of the spin polarization vector $|\vec{S}|$ for states in the two conduction bands obtained with the 4-bands low-energy model. The results correspond to adatom concentration 8 % (1/4 ML) (*left pannel*) and 25 % (*right pannel*). In both cases, close to the Dirac point, spin and pseudospin entanglement is very high given the small values of $|\vec{S}| \ll 1$

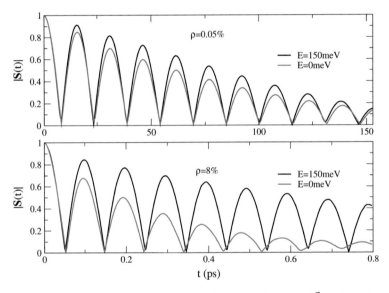

Fig. 5.7 Time-dependence of the modulus of the spin polarization vector $|\vec{S}(E, t)|$ in the microscopic model with realistic disorder and gold ad-atom concentrations 0.05 and 8 % at two specific energies: Dirac point $E = 0$ and $E = 150$ meV

Energy crossover of spin/pseudospin dynamics and effective magnetic/pseudomagnetic fields

Figure 5.4 exhibits different oscillating periods for spin and pseudospin. At high energy, spin precession leads to oscillations in $\vec{S}(t)$ with Rashba period T_Ω while the pseudospin oscillations ($\langle \sigma_z(t) \rangle$ in Fig. 5.4) are driven by T_Ω together with a more rapid superimposed oscillation. A crossover to complex low-energy dynamics is observed where spin- and pseudospin motion are more closely related to one another. To illustrate the relation between spin, pseudospin and momentum, we introduce three different effective pseudomagnetic fields:

$$
\begin{aligned}
\vec{B}_0^{\mathrm{ps}}(\vec{k}) &= \hbar v_F(\eta k_x, k_y, 0) \\
\vec{B}_R^{\mathrm{ps}}(\vec{s}) &= \overline{\lambda}_R(\eta\langle s_y\rangle, -\langle s_x\rangle, 0) \\
\vec{B}_I^{\mathrm{ps}}(\vec{s}) &= \overline{\lambda}_I(0, 0, \eta\langle s_z\rangle)
\end{aligned}
\tag{5.8}
$$

and two effective magnetic fields which are extracted from Eq. (5.3):

$$
\begin{aligned}
\vec{B}_R(\vec{\sigma}) &= \overline{\lambda}_R(-\langle\sigma_y\rangle, \eta\langle\sigma_x\rangle, 0) \\
\vec{B}_I(\vec{\sigma}) &= \overline{\lambda}_I(0, 0, \eta\langle\sigma_z\rangle)
\end{aligned}
\tag{5.9}
$$

where $\langle\sigma_i\rangle = \langle\Psi_{\vec{k}}|\sigma_i\otimes 1_s|\Psi_{\vec{k}}\rangle$ and $\langle s_i\rangle = \langle\Psi_{\vec{k}}|s_i\otimes 1_\sigma|\Psi_{\vec{k}}\rangle$ are the expectation values of the pseudospin and spin operators, σ_i and s_i (analogous to decoupled subsystems), and $\Psi_{\vec{k}}$ are the eigenstates of the KMR Hamiltonian (see Eq. (5.5)). From the form of the effective magnetic fields, it is seen that unlike the case of semiconductors where SOC directly couples spin with momentum, in graphene spin couples directly with pseudospin (see B_R and B_R^{ps}), and is related to momentum via the coupling between pseudospin and momentum $h_0(\vec{k})$, a term which vanishes at the Dirac point. These effective magnetic fields help illustrating the energy crossover in Fig. 5.4.

While the occurrence of the same Rashba precession for spin and pseudospin at high energy (Fig. 5.4a) is related to the analogy of the effective fields (B_R for spin and B_R^{ps} for pseudospin), the superimposed rapid oscillation in σ_z can be rationalized as follows. We observe that at high energy the nearest neighbor hopping from three neighbors, $h_0 \propto k$, dictates additional pseudospin precession about a radial in-plane field $B_0^{\mathrm{ps}}(\propto k) = h v_F(k_x, k_y, 0)$ with small amplitudes for $\langle\sigma_z(t)\rangle$ and with a period given by $T_0^{\mathrm{ps}} = \pi\hbar/E$ (0.016 ps for $E = 130$ meV). For the overall dynamics it is important that this rapid pseudospin precession about B_0^{ps} does not affect the slower spin dynamics imposed by h_R. Indeed we can replace σ_x by its time average $\overline{\langle\sigma_x\rangle}$ and $\sigma_y \to \overline{\langle\sigma_y\rangle}$ in B_R. As a result, there is only weak interference (feedback) between spin and pseudospin dynamics and both degrees of freedom can be understood as being driven independently by their respective effective fields.

In contrast, at low energy, the above replacements are no longer justified and B_R becomes time dependent through the time dependence of $\langle\vec{\sigma}\rangle$ (analogously for B_R^{ps} and $\langle\vec{s}\rangle$) resulting in complex spin-pseudospin dynamics with new characteristic periods.

Momentum relaxation, spin relaxation and entanglement of states in gold-decorated graphene samples

From the analysis of spin dynamics using the microscopic and continuum models (Figs. 5.3 and 5.4), we have shown that the spin relaxation mechanism close to the Dirac point is inconsistent with EY or DP scaling laws. For EY, the spin relaxation time is proportional to the momentum relaxation time as $\tau_s^{EY} \approx N_{collisions}\cdot\tau_p$, where

$N_{collisions} \gg 1$ denotes the number of scattering-off-impurity events before spin flip occurs and τ_p is the transport time. By definition $\tau_s^{EY} \gg \tau_p$ which is opposite to our estimates in the low impurity regime.

For the DP mechanism, the scaling behavior between spin and momentum relaxation times is inverted $\tau_s^{DP} \propto 1/\tau_p$. The essential characteristic of such mechanism is however that if disorder increases (accompanied by a decay of τ_p) then τ_s increases consistently. Our results cannot be described by such scaling since by increasing disorder μ, both τ_p and τ_s decrease simultaneously. Also, when approaching the Dirac point, τ_p seems to increase continuously while τ_s tends to saturate to a minimum finite value. This is similarly seen in the time-dependence of the modulus of the spin polarization vector $|\vec{S}(t)|$ in the microscopic model with realistic disorder. Figure 5.7 shows $|\vec{S}(t)|$ for 0.05 and 8 % gold ad-atom concentrations and are complementary to Fig. 5.2 of the main paper in showing the spin polarization loss accumulated in time [30]. The fact that the total spin polarization decreases faster when approaching the Dirac point, where momentum relaxation time (τ_p) is larger, is a further confirmation of our interpretation that spin-pseudospin entanglement driven by Rashba-type SOC is at the heart of the spin relaxation mechanism of gold-decorated graphene at low energies.

Influence of charge puddles.

It is well known that charge puddles in graphene can make the Dirac point energy fluctuating due to local changes in the chemical potential [31] possibly hindering the observation of the discussed spin-relaxation mechanism at low energy. As recently reported by Xue et al. [32], the fluctuation (standard deviation) of the Dirac point energy in supported graphene samples depends on the substrate and range from $\Delta E \approx 56$ meV for SiO_2 to $\Delta E \approx 5$ meV for hBN [32] in consistency with a previous paper reporting 50 meV for SiO_2 [33]. The difference in the energetic position of the Dirac point is intimately related to the size of the charge puddles induced by the substrate which reach an approximate size of 10 nm for SiO_2 and around 100 nm for hBN. By comparing the relevant energies for the band onsets (i.e. energies where $|\vec{S}| \ll 1$) in Fig. 5.6 and the Fermi energy fluctuations from literature, we expect that the spin relaxation mechanism proposed in this manuscript, based on spin-pseudospin entanglement, should be experimentally accessible.

Analogy to spin-less bilayer graphene We observe that the Hamiltonian in Eq. (5.3) is very similar to the one of spin-less bilayer graphene (BLG) at low energies and shows a very similar band dispersion around both valleys [34–36], although the nature of eigenstates is quite different [12]. Below we compare the Hamiltonian matrices of both cases. The KMR Hamiltonian in one valley reads

$$\mathcal{H}_{KMR}^{(K)} = \begin{pmatrix} \lambda_I & v(k_x - ik_y) & 0 & 0 \\ v(k_x + ik_y) & -\lambda_I & -2i\lambda_R & 0 \\ 0 & 2i\lambda_R & -\lambda_I & v(k_x - ik_y) \\ 0 & 0 & v(k_x + ik_y) & \lambda_I \end{pmatrix} \quad (5.10)$$

and the spin-less BLG-Hamiltonian, in its most reduced version [36], can be expressed as:

$$
\mathcal{H}_{BLG}^{(K)} = \begin{pmatrix} -\Delta' & v(k_x - ik_y) & 0 & 0 \\ v(k_x + ik_y) & \Delta' & \gamma_1 & 0 \\ 0 & \gamma_1 & \Delta' & v(k_x - ik_y) \\ 0 & 0 & v(k_x + ik_y) & -\Delta' \end{pmatrix} \tag{5.11}
$$

where γ_1 is the interlayer hopping which connects a B-site in the top layer with an A-site in the bottom layer in Bernal stacked bilayer graphene. This interaction induces a staggered potential $\pm\Delta'$ within each layer distinguishing carbon atoms in top position and those at hollow-sites. Interestingly, this staggered potential changes sign at opposite layers similarly to the intrinsic SOC λ_I in graphene.

It is helpful to write the above Hamiltonian in terms of Pauli matrices in order to compare with Eq. (5.3):

$$
\begin{aligned}
h_0^{BLG}(\vec{k}) &= \hbar v_F(\eta \sigma_x k_x + \sigma_y k_y) \otimes 1_s \\
h_\gamma^{BLG}(\vec{k}) &= \gamma_1 \left([\sigma_x \otimes \xi_x] + [\sigma_y \otimes \xi_y]\right) \\
h_\Delta^{BLG}(\vec{k}) &= \Delta'\left[\sigma_z \otimes \xi_z\right]
\end{aligned} \tag{5.12}
$$

where the layer operator $\vec{\xi}$ in Eq. (5.12) plays the role of the spin operator $\vec{s}$ in Eq. (5.3), while the second degree of freedom is the pseudospin $\vec{\sigma}$ in both cases. It is important to note that, while the first and third terms in Eq. (5.12) resemble the ones in Eq. (5.3), the second one has a different structure in terms of Pauli matrices when compared to the Rashba term. However, it also leads to in-plane effective pseudomagnetic and magnetic fields of the form:

$$
\vec{B}_\gamma^{ps}(\vec{\xi}) = \gamma_1(\langle \xi_x \rangle, \langle \xi_y \rangle, 0) \tag{5.13}
$$

$$
\vec{B}_\gamma(\vec{\sigma}) = \gamma_1(\langle \sigma_x \rangle, \langle \sigma_y \rangle, 0). \tag{5.14}
$$

Also, the eigenstates of the BLG Hamiltonian, while no longer complex, still show a layer-pseudospin entanglement at low-energies allowing for new interesting phenomena regarding layer relaxation in Bernal stacked bilayer graphene

$$
\Psi_{\vec{k},\pm}^{BLG,I} = \begin{pmatrix} 0 \\ 1 \end{pmatrix} \otimes |1\rangle \pm \begin{pmatrix} 1 \\ 0 \end{pmatrix} \otimes |2\rangle \tag{5.15}
$$

$$
\Psi_{\vec{k},\pm}^{BLG,II} = \begin{pmatrix} 1 \\ 0 \end{pmatrix} \otimes |1\rangle \pm \begin{pmatrix} 0 \\ 1 \end{pmatrix} \otimes |2\rangle. \tag{5.16}
$$

The apparent similarity of both Hamiltonians indicates the possibility to observe physical effects similar to the presently studied spin relaxation and spin-pseudospin entanglement when considering 'layer-polarized' carrier transport in graphene. It would be interesting to study the effect of layer-pseudospin entanglement in such a situation.

In conclusion, our spin transport simulations in graphene, chemically modified by a random distribution of ad-atoms, have revealed a hitherto unknown phenomenon related to the entangled dynamics of spin and pseudospin, which is induced by SOC and leads to fast spin relaxation in a quasi-ballistic transport regime. The entanglement between spin and orbital degrees of freedom has been discussed for models of ballistic semiconducting nanowires [30]. Here, the energy-dependence of spin/pseudospin entanglement induced by SOC has been shown to directly impact the resulting spin dynamics and spin relaxation times. Faster spin relaxation develops when spin-pseudospin entanglement is maximized at the Dirac point, where the momentum scattering time becomes increasingly large because disorder preserves pseudospin symmetry.

This relaxation mechanism, occurring in clean graphene with long mean free paths, has no equivalent in condensed matter and cannot be described by EY or DP scaling. Such a phenomenon is here revealed for the specific case of gold adsorbates, but should also be at play for other sources of local SOC (ripples, defects, etc.), thus contributing to a deep general understanding of spin transport in graphene-based materials and devices [1, 16–18, 37], while the specific spin relaxation time depends on the effective strength of the SOC being different for different sources. The effect of lateral confinement in stripe or ribbon geometry deserves further investigation regarding its influence on spin relaxation (which was observed in semiconductor nanowires [38]), while some general mechanism due to flexural phonons for spin relaxation in 2D membranes has been proposed [39].

Finally, the spin-pseudospin entanglement could open the path to control the pseudospin by modifying the spin or vice versa. For example, spins could be manipulated by inducing pseudomagnetic fields by straining graphene. Such possibilities could lead to the development of novel approaches for non-charge-based information processing and computing, resulting in a new generation of active (CMOS-compatible) spintronic devices together with non-volatile low-energy MRAM memories [40].

5.2 Quantum Spin Hall Effect

5.2.1 Introduction

In 2005, Kane and Mele predicted the existence of the Quantum Spin Hall Effect (QSHE) in graphene due to intrinsic SOC [41, 42]. Within the QSHE, the presence of SOC, which can be understood as a momentum-dependent magnetic field coupling to the spin of the electron, results in the formation of chiral (anti-chiral) integer QHE for spin up (spin down) electron population. The observation of QSHE has been however prohibited in clean graphene by the vanishingly small intrinsic SOC in the order of μeV [43], but further realized in strong SOC materials (such as CdTe/HgTe/CdTe quantum wells or bismuth selenide and telluride alloys), giving rise to the new exciting field of topological insulators [44–47]. Recent proposals to induce a topological phase in graphene include functionalization with heavy adatoms [25, 48], covalent functionalization of the edges [49], proximity effect with other topological insulators [50–52], or intercalation and functionalization with 5d transition metals [53, 54]. In particular, the seminal theoretical study [25] by Weeks and co-workers has revealed that graphene endowed with modest coverage of heavy adatoms (such as indium and thallium) could exhibit a substantial band gap and QSHE fingerprints (detectable in transport or spectroscopic measurements). For instance, signature of such a topological state could be seen in a robust quantized two-terminal conductance ($2e^2/h$), with an adatom density dependent conductance plateau extending inside the bulk gap induced by SOC [25, 55, 56]. To date, such a prediction lacks experimental confirmation, despite some recent results on indium-functionalized graphene have shown a surprising reduction of the Dirac point resistance with increasing indium density [57]. On the other hand, it is known that adatoms deposited on graphene will inevitably segregate, forming islands rather than a homogeneous distribution [58]. Such a clustering effect may seriously impact on the transport features [59–61].

In this Letter, we show that the clustering of thallium adatoms on graphene could suppress the formation of a quantum spin-Hall phase, while the resulting functionalized structures would exhibit unconventional bulk transport characteristics, with absence of transition to an insulating regime and a robust Dirac point conductivity close to $4e^2/h$. The presence of adatom islands locally introducing strong SOC is actually found to prevent the development of quantum interferences and localization phenomena induced by additional strong disorder sources.

5.2.2 Adatom Clustering Effect on QSHE

Model and Methods. When a thallium atom is grafted on graphene, it places in the middle of a hexagonal plaquette of carbon atoms, above the surface, see Fig. 5.8. As shown in [25], the degrees of freedom corresponding to the adatom can be conveniently decimated and their effect included into an effective π-π^* orthogonal TB model with SOC. In the presence of adatoms randomly distributed over a set $\mathcal{R}$ of plaquettes, the Hamiltonian [41, 42] reads as

$$\hat{\mathcal{H}} = -\gamma_0 \sum_{\langle ij \rangle} c_i^\dagger c_j + \frac{2i}{\sqrt{3}}\lambda \sum_{\langle\langle ij \rangle\rangle \in \mathcal{R}} c_i^\dagger \vec{s} \cdot (\vec{d}_{kj} \times \vec{d}_{ik}) c_j$$
$$- \mu \sum_{i \in \mathcal{R}} c_i^\dagger c_i + \sum_i V_i c_i^\dagger c_i , \qquad (5.17)$$

where $c_i = [c_{i\downarrow}, c_{i\uparrow}]$ is the couple of annihilation operators for electrons with spin down and spin up on the ith carbon atom, and $c_i^\dagger$ is the corresponding couple of

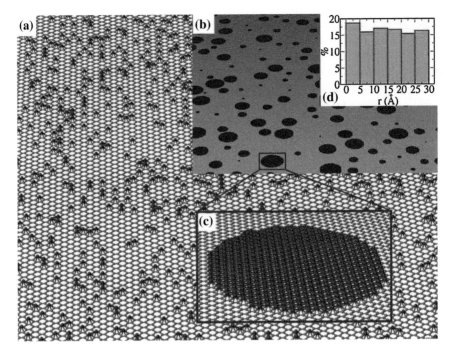

Fig. 5.8 a Ball-and-stick model of a graphene substrate with randomly adsorbed thallium atoms (concentration is 15 %). **b** Same as (**a**) but with adatoms clustered in islands with a radius distribution varying up to 3 nm (histogram shown in (**d**)). **c** Zoom-in of a typical thallium ad-atoms-based island. All thallium atoms are positioned in the hollow position and equally connected to the 6 carbon atoms forming the hexagon underneath (following [25])

creation operators. The first contribution in Eq. (5.17) is the nearest neighbor hopping TB term, with coupling energy $\gamma_0 = 2.7\,\text{eV}$. The second contribution is a next nearest neighbor hopping term that represents the SOC induced by the adatoms, with $\vec{d}_{kj}$ and $\vec{d}_{ik}$ the unit vectors along the two bonds connecting second neighbors and $\vec{s}$ the spin Pauli matrices. The SOC is set to $\lambda = 0.02\gamma_0$, as extracted from *ab-initio* simulations in Ref. [25]. The third term describes the potential energy induced by charge transfer between adatoms and graphene. The last term represents the long-range interaction of graphene and impurities in the substrate $V_i = \sum_{j=1}^{N} \epsilon_j \exp[-(\mathbf{r}_i - \mathbf{R}_j)^2/(2\xi^2)]$ [62], where $\xi = 0.426$ nm is the effective range and the sum runs over N impurity centers with random positions $\mathbf{R}_j$ and magnitude of the potential ϵ_j randomly chosen within $[-\Delta, \Delta]$. The Hamiltonian does not consider the effect of a further structure relaxation in the case of clustered adatoms. This will not alter our conclusions.

For the study of electronic transport in thallium-functionalized ribbons, we consider a standard two-terminal configuration with highly doped contacts. The doping is mimicked by an appropriate potential energy V on source and drain. The simulations are based on the nonequilibrium Green's function formalism [63]. In addition to the electronic conductance, this approach provides us with the spin-resolved local density-of-occupied-states. This quantity illustrates how electrons injected from the source spatially distribute in the system depending on their spin. More specifically, the zero-temperature differential conductance as a function of the electron energy is obtained by the Landauer-Büttiker formula

$$G(E) = (e^2/h)\text{Tr}[G^R(E)\Gamma^{(\text{S})}G^A(E)\Gamma^{(\text{D})}]\,, \qquad (5.18)$$

where $G^{R/A}$ are the retarded and advanced Green's functions and $\Gamma^{(\text{S/D})}$ are the rate operators for the source and drain contacts. The local density-of-occupied-states is obtained as

$$\rho_{i\eta}(E) = \Im m[G^<(E)]_{i\eta,i\eta}/(2\pi) \qquad (5.19)$$

where $[G^<]_{i\eta,i\eta}$ is the diagonal element of the lesser Green's function corresponding to the electron with spin η ($\downarrow$, $\uparrow$) of the ith carbon atom, and $\Im m$ indicates the imaginary part.

We also study quantum transport in two-dimensional functionalized graphene by means of the Kubo approach [29, 64]. The scaling properties of the conductivity can be followed through the dynamics of electronic wavepackets using Eq. (3.41). Calculations, based on the use of Chebyshev polynomial expansion and continued fractions, are performed on systems containing more than 3.5 million carbon atoms, which corresponds to sizes larger than $300 \times 300\,\text{nm}^2$. Such a size guarantees that our results are weakly dependent on the specific spatial distribution of adatoms or clusters of adatoms.

Suppression of QSHE by adatom clustering. We start by considering an armchair ribbon of width $W = 50\,\text{nm}$ functionalized with a concentration $n = 15\%$ of randomly scattered thallium adatoms over a length $L = 50\,\text{nm}$. As already reported in the literature [25], the differential conductance [continuous line in Fig. 5.9a] clearly shows a $2e^2/h$ plateau, which is signature of quantum spin-Hall phase. Note that the plateau is centered at $E \approx -120$ meV and has an extension of about 100 meV. The observed charge neutrality point shift is consistent with the concentration of carbon atoms $\sim 3n$ that undergo a charge transfer doping effect, i.e. $E \approx -3n\mu = -121.5$ meV. The width of the plateau approximately corresponds to the topological gap induced by thallium functionalization, and is given by $6\sqrt{3}\lambda_{\text{eff}} \approx 84.2$ meV, where the effective SOC is $\lambda_{\text{eff}} = n\lambda \approx 8.1$ meV [65]. A closer inspection of the conductance shows that actually the plateau region is not perfectly flat, but varies within the range [1.92,2.02] e^2/h. This indicates that the separation between spin polarized chiral edge channels is not complete. A better quantization may be achieved by increasing W, L or the adatom concentration. Figures 5.9b, c also show the spin resolved local density-of-occupied-states ρ for electrons injected from the right contact at energy $E = -100$ meV, indicated by an arrow in Fig. 5.9a. We observe a high ρ for $x > 50$ nm, i.e. the region of the source (injected electrons are indicated by arrows), and spin-polarized channels along the upper edge for spin down (b) and along the lower edge for spin up (c). The width of the polarized edge channels in armchair ribbons does not depend on the energy but only on the SOC as $a\gamma_0/(2\sqrt{3}\lambda_{\text{eff}}) \approx 13.5$ nm (see [66]). The separation between the right-to-left and

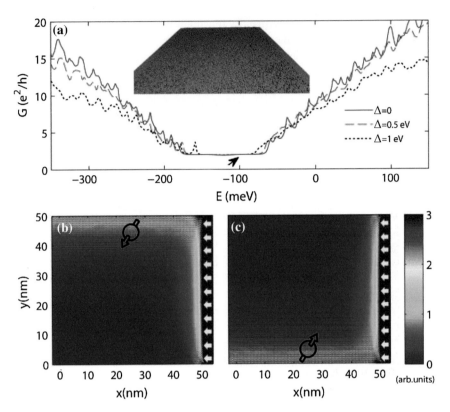

Fig. 5.9 a Differential conductance for an armchair ribbon of width $W = 50$ nm with a concentration $n = 15\%$ of randomly scattered thallium adatoms over a section with length $L = 50$ nm. The potential energy on the contacts is set to $V = -2.5$ eV. The presence of long-range disorder with Δ up to 1 eV is taken into account. **b** Local density-of-occupied-states for spin down electrons injected from the right contact for $\Delta = 0$ at energy $E = -100$ meV, see the arrow in (**a**). **c** Same as (**b**) but for spin up electrons

left-to-right moving channels, which is opposite for different spin polarizations, is at the origin of the QSHE. To test its robustness, we consider the presence of a concentration $n_{LR} = 0.5\%$ of long-range disorder with different strength Δ. As reported in Fig. 5.9a, a plateau, though narrower, is observed up to $\Delta = 1$ eV.

This picture is actually strongly modified when adatoms segregate and form islands. Figure 5.10a shows the evolution of the differential conductance when islands have a radius r varying from 0 (non-segregated case) to 1 nm and finally to random values between 2 and 3 nm. The adatom concentration is kept at $n = 15\%$. While a signature of the plateau remains up to $r = 1$, for larger radius the quantization is completely lost despite the short intercluster distance. This indicates that *segregation has a detrimental effect on the formation of a QSH phase in graphene by heavy adatom functionalization.* Considering that adatom clustering is unavoidable at room temperature, our findings provide an explanation for the missing experimental observation

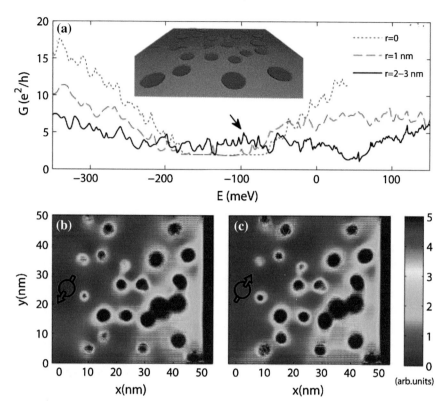

Fig. 5.10 **a** Differential conductance for an armchair ribbon of width $W = 50$ nm with a concentration $n = 15\%$ of clustered thallium adatoms (in islands with radius r up to 2–3 nm) over a section with length $L = 50$ nm. The potential energy on the contacts is set to $V = -2.5$ eV. **b** Local density-of-occupied-states in the case $r \in [2, 3]$ nm, for spin down electrons injected from the right contact at energy $E = -100$ meV, see the arrow in (**a**). **c** Same as (**b**) but for spin up electrons

of the QSHE in such systems. A deeper insight into the effect of segregation is further provided by the spin-resolved density-of-occupied-states reported in Figs.5.10b, c for the case of island radii in the range [2, 3] nm. The ρ distribution is very similar for spin down and spin up electrons, this means that most of the spin-coupling related effect is suppressed. Moreover, the injected electrons largely spread all over the ribbon and show a higher concentration inside the islands. To explain these features, we have to consider that segregation reduces the homogeneous coverage of adatoms and leaves large regions of pristine graphene. As a consequence, the topological gap cannot develop in these regions, where electrons flow the same way as in non-topological systems. Moreover, the clusters are too small and the SOC is too weak to induce a topological phase inside them. Together with the highly negative value of the charge neutrality point inside the islands ($E = 3\mu = -810$ meV), this determines the considerably high electron density observed in the figure. However, as shown below, clustering of thallium adatoms produces a remarkable bulk transport fingerprint of the SOC in two-dimensional graphene.

Robust metallic state and minimum conductivity. We investigate the intrinsic bulk conductivity of thallium-functionalized graphene by computing the Kubo-Greenwood conductivity. We focus on large thallium density (about 15%), with thallium clusters size distribution shown in Fig. 5.8d and consider superimposed distribution of long-range impurities to mimic additional sources of disorder (such as charged defects trapped in the underneath oxide, additional dopants, structural defects...). In Fig. 5.11 (main frame), we show the Kubo conductivity for various densities ($n_{LR} = 0.2 - 0.5\%$) of long-range impurities with $\Delta = 2.7$ eV. A striking feature is the energy-dependent impact of additional disorder on the transport features. Indeed a plateau is formed near the Dirac point, where the conductivity reaches a minimum value, regardless of the superimposed disorder potential. Differently, a more conventional scaling behavior $\sigma \sim 1/n_{LR}$ is obtained for high energies, following a semiclassical Fermi golden rule. The minimum conductivity obtained $\sigma_{min} \sim 4e^2/h$ reminds the case of clean graphene deposited on oxide substrates and sensitive to electron-hole puddles [67]. However, here the role of spin-orbit interaction is critical for preserving a robust metallic state. This is shown in Fig. 5.11 (inset), where the time-dependence of the diffusion coefficient at energy ($E = -120$ meV) is reported for $n_{LR} = 0.5\%$, in presence of the thallium islands with and without spin-orbit interaction. The absence of SOC irremediably produces an insulating state as evidenced by the decay of the diffusion coefficient, whereas once SOC is switched on, the diffusivity is found to saturate to its semiclassical values, showing no sign of quantum interferences and localization, in agreement with a percolation scenario for the corresponding electronic states. Note that such a mechanism is not

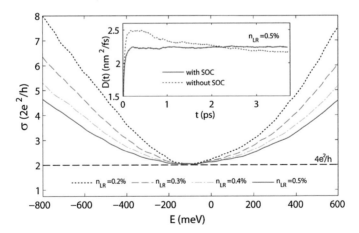

Fig. 5.11 Kubo conductivity versus energy for thallium clustering and additional varying density (n_{LR}) of long-range impurities. *Inset* Diffusion coefficient for wavepacket with energy $E = -120$ meV, for the case $n_{LR} = 0.5\%$, with (*solid blue line*) and without the SOC of thallium adatoms activated

connected with WAL (a phenomenon that has been studied in graphene in presence and absence of SOC [8, 62, 68–70]) since the origin of quantum interferences effects is disconnected from the local contribution of SOC underneath the formed islands. This highlights a *residual bulk signature of the SOC in the diffusive transport regime at the Dirac point*.

Conclusion. We have theoretically shown and quantified the detrimental effect of heavy adatom clustering on the formation of the QSHE phase. An inhomogeneous surface coverage by adatom quenches the topological gap and the formation of the topologically protected spin-polarized edge transport channels. Simultaneously, the intrinsic bulk conductivity reveals peculiar features, such as the absence of localization and a robust minimum conductivity in the vicinity of the Dirac point, resulting from a percolation of propagating states between islands. Those findings might guide future experiments on the way to fabricate and realize a topological insulating phase based on chemically functionalized graphene [57]. Furthermore, our prediction of an unconventional bulk metallic phase, once the has been suppressed, opens new venues for exploring new and original spin-orbit related quantum transport phenomena in graphene/topological-insulator hybrid systems [52]. In the same perspective, it would be interesting to study the formation and robustness of a quantum anomalous Hall phase in presence of segregation of $3d$ transition metallic adatoms on graphene [55, 71].

References

1. N. Tombros et al., Nature (London) **448**, 571 (2007)
2. N. Tombros et al., Phys. Rev. Lett. **101**, 046601 (2008)
3. C. Jozsa et al., Phys. Rev. Lett. **100**, 236603 (2008)
4. T. Maassen et al., Nano Lett. **12**, 1498–1502 (2012)
5. D. Huertas-Hernando, F. Guinea, A. Brataas, Phys. Rev. Lett. **103**, 146801 (2009)
6. A.H. Castro Neto, F. Guinea, N.M.R. Peres, K.S. Novoselov, A.K. Geim, Rev. Mod. Phys. **81**, 109 (2009)
7. M.I. Katsnelson, K.S. Novoselov, A.K. Geim, Nat. Phys. **2**, 620 (2006)
8. E. McCann, K. Kechedzhi, V.I. Falko, H. Suzuura, T. Ando, B.L. Altshuler, Phys. Rev. Lett. **97**, 146805 (2006)
9. K.S. Novoselov, D. Jiang, Y. Zhang, S.V. Morozov, H.L. Stormer, U. Zeitler, J.C. Maan, G.S. Boebinger, P. Kim, A.K. Geim, Science **315**, 1379 (2007)
10. A. Rycerz, J. Tworzydo, C.W.J. Beenakker, Nat. Phys. **3**, 172–175 (2007)
11. P. San-Jose, E. Prada, E. McCann, H. Schomerus, Phys. Rev. Lett. **102**, 247204 (2009)
12. E.I. Rashba, Phys. Rev. B **79**, 161409(R) (2009)
13. I. Zutic, J. Fabian, S. Das Sarma, Rev. Mod. Phys. **76**, 323 (2004)
14. F.J. Jedema, H.B. Heersche, A.T. Filip, J.J.A. Baselmans, B.J. van Wees, Nature **416**, 713–716 (2002)
15. X. Lou, C. Adelmann, S.A. Crooker, E.S. Garlid, J. Zhang, K.S. Madhukar Reddy, S.D. Flexner, C.J. Palmstrom, P.A. Crowell, Nat. Phys. **3**, 197–202 (2007)
16. A. Avsar et al., Nano Lett. **11**, 2363 2368 (2011)
17. W. Han, R.K. Kawakami, Phys. Rev. Lett. **107**, 047207 (2011)
18. P.J. Zomer, M.H.D. Guimaraes, N. Tombros, B.J. van Wees, Phys. Rev. B **86**, 161416(R) (2012)
19. H. Ochoa, A.H. Castro Neto, F. Guinea, Phys. Rev. Lett. **108**, 206808 (2012)

20. Z. Zhang, M.W. Wu, New J. Phys. **14**, 033015 (2012)
21. C. Ertler et al., Phys. Rev. B **80**, 041405 (2009)
22. A.H. Castro Neto, F. Guinea, Phys. Rev. Lett. **103**, 026804 (2009)
23. M.H.D. Guimaraes, A. Veligura, P.J. Zomer, T. Maassen, I.J. Vera-Marun, N. Tombros, B.J. van Wees, Nano Lett. **12**, 3512–3517 (2012)
24. I. Neumann, J. Van de Vondel, G. Bridoux, M.V. Costache, F. Alzina, C.M. Sotomayor Torres, S.O. Valenzuela, Small **9**, 156–160 (2013)
25. C. Weeks, J. Hu, J. Alicea, M. Franz, R. Wu, Phys. Rev. X **1**, 021001 (2011)
26. Yu.S. Dedkov, M. Fonin, U. Rudiger, C. Laubschat, Phys. Rev. Lett. **100**, 107602 (2008)
27. D. Marchenko, A. Varykhalov, M.R. Scholz, G. Bihlmayer, E.I. Rashba, A. Rybkin, A.M. Shikin, O. Rader, Nat. Commun. **3**, 1232 (2012)
28. K. Pi et al., Phys. Rev. Lett. **104**, 187201 (2010)
29. S. Roche, N. Leconte, F. Ortmann, A. Lherbier, D. Soriano, J.-C. Charlier, Solid State Commun. **152**, 1404–1410 (2012)
30. B.K. Nikolic, S. Souma, Phys. Rev. B **71**, 195328 (2005)
31. A. Deshpande, W. Bao, F. Miao, C.N. Lau, B.J. LeRoy, Phys. Rev. B **79**, 205411 (2009)
32. J. Xue et al., Nat. Phys. **10**, 282–285 (2011)
33. J. Martin, N. Akerman, G. Ulbright, T. Lohmann, J.H. Smet, K. von Klitzing, A. Yacoby, Nat. Phys. **4**, 144 (2008)
34. F. Guinea, A.H. Castro-Neto, N.M.R. Peres, Phys. Rev. B **73**, 245426 (2006)
35. E. McCann, V.I. Falko, Phys. Rev. Lett. **96**, 086805 (2006)
36. E. McCann, M. Koshino, Rep. Prog. Phys. **76**, 056503 (2013)
37. B. Dlubak, M.-B. Martin, C. Deranlot, B. Servet, S. Xavier, R. Mattana, M. Sprinkle, C. Berger, W.A. De Heer, F. Petroff, A. Anane, P. Seneor, A. Fert, Nat. Phys. **8**, 557 (2012)
38. A.W. Holleitner, V. Sih, R.C. Myers, A.C. Gossard, D.D. Awschalom, Phys. Rev. Lett. **97**, 036805 (2006)
39. H. Dery, Y. Song, Phys. Rev. Lett. **111**, 026601 (2013)
40. H. Dery, H. Wu, B. Ciftcioglu, M. Huang, Y. Song, R. Kawakami, J. Shi, I. Krivorotov, I. Zutic, L.J. Sham, IEEE Trans. Electron Devices **59**, 259–262 (2012)
41. C.L. Kane, E.J. Mele, Phys. Rev. Lett. **95**, 226801 (2005)
42. C.L. Kane, E.J. Mele, Phys. Rev. Lett. **95**, 146802 (2005)
43. Y. Yao et al., Phys. Rev. B **75**, 041401(R) (2007)
44. B.A. Bernevig, S.-C. Zhang, Phys. Rev. Lett. **96**, 106802 (2006)
45. B.A. Bernevig, T.L. Hughes, S.-C. Zhang, Science **314**, 1757 (2006)
46. M.Z. Hasan, C.L. Kane, Rev. Mod. Phys. **82**, 3045 (2010)
47. X.-L. Qi, S.-C. Zhang, Rev. Mod. Phys. **83**, 1057 (2011)
48. H. Jiang, Z. Qiao, H. Liu, J. Shi, Q. Niu, Phys. Rev. Lett. **109**, 116803 (2012)
49. G. Autes, O.V. Yazyev, Phys. Rev. B **87**, 241404 (2013)
50. K.-H. Jin, S.-H. Jhi, Phys. Rev. B **87**, 075442 (2013)
51. W. Liu, X. Peng, X. Wei, H. Yang, G.M. Stocks, J. Zhong, Phys. Rev. B **87**, 205315 (2013)
52. L. Kou, B. Yan, F. Hu, S.-C. Wu, T.O. Wehling, C. Felser, C. Chen, T. Frauenheim, Nano Lett. **13**, 6251 (2013)
53. J. Hu, J. Alicea, R. Wu, M. Franz, Phys. Rev. Lett. **109**, 266801 (2012)
54. Y. Li, P. Tang, P. Chen, J. Wu, B.-L. Gu, Y. Fang, S.B. Zhang, W. Duan, Phys. Rev. B **87**, 245127 (2013)
55. Z. Qiao, S.A. Yang, W. Feng, W.-K. Tse, J. Ding, Y. Yao, J. Wang, Q. Niu, Phys. Rev. B **82**, 161414 (2010)
56. Z. Qiao, W.-K. Tse, H. Jiang, Y. Yao, Q. Niu, Phys. Rev. Lett. **107**, 256801 (2011)
57. J. Coraux, L. Marty, N. Bendiab, V. Bouchiat, Acc. Chem. Res. **46**, 2193 (2013)
58. E. Sutter, P. Albrecht, B. Wang, M.-L. Bocquet, L. Wu, Y. Zhu, P. Sutter, Surf. Sci. **605**, 1676 (2011)
59. K.M. McCreary, K. Pi, A.G. Swartz, W. Han, W. Bao, C.N. Lau, F. Guinea, M.I. Katsnelson, R.K. Kawakami, Phys. Rev. B **81**, 115453 (2010)

60. M. Alemani, A. Barfuss, B. Geng, C. Girit, P. Reisenauer, M.F. Crommie, F. Wang, A. Zettl, F. Hellman, Phys. Rev. B **86**, 075433 (2012)
61. T. Eelbo, M. Wasniowska, P. Thakur, M. Gyamfi, B. Sachs, T.O. Wehling, S. Forti, U. Starke, C. Tieg, A.I. Lichtenstein, R. Wiesendanger, Phys. Rev. Lett. **110**, 136804 (2013)
62. F. Ortmann, A. Cresti, G. Montambaux, S. Roche, Europhys. Lett. **94**, 47006 (2011)
63. A. Cresti, G. Grosso, G. Pastori Parravicini, Eur. Phys. J. B **53**, 537 (2006)
64. L.E.F. Foa Torres, S. Roche, J.C. Charlier, *Introduction to Graphene-Based Nanomaterials From Electronic Structure to Quantum Transport* (Cambridge, 2013)
65. O. Shevtsov, P. Carmier, C. Groth, X. Waintal, D. Carpentier, Phys. Rev. B **85**, 245441 (2012)
66. E. Prada, G. Metalidis, J. Comput. Electron. **12**, 63 (2013)
67. S.D. Sarma, S. Adam, E.H. Hwang, E. Rossi, Rev. Mod. Phys. **83**, 407 (2011)
68. E. McCann, V.I. Falko, Phys. Rev. Lett. **108**, 166606 (2012)
69. A. Rycerz, J. Tworzydo, C.W.J. Beenakker, Europhys. Lett. **79**, 57003 (2007)
70. Y.Y. Zhang, J. Hu, B.A. Bernevig, X.R. Wan, X.C. Xie, W.M. Liu, Phys. Rev. Lett. **102**, 106401 (2009)
71. J. Ding, Z. Qiao, W. Feng, Y. Yao, Q. Niu, Phys. Rev. B **84**, 195444 (2011)

Chapter 6
Conclusions

In this thesis, I have presented the charge transport of disordered graphene as well as explained the fast spin relaxation in graphene which is one of the most interesting topics in graphene at the moment.

The role of defect-induced zero-energy modes on charge transport in graphene is investigated using Kubo and Landauer transport calculations. By tuning the density of random distributions of monovacancies either equally populating the two sublattices or exclusively located on a single sublattice, all conduction regimes are covered from direct tunneling through evanescent modes to mesoscopic transport in bulk disordered graphene. Depending on the transport measurement geometry, defect density, and broken sublattice symmetry, the Dirac-point conductivity is either exceptionally robust against disorder (supermetallic state) or suppressed through a gap opening or by algebraic localization of zero-energy modes, whereas weak localization and the Anderson insulating regime are obtained for higher energies. These findings clarify the contribution of zero-energy modes to transport at the Dirac point, hitherto controversial.

We also reported new insights to the current understanding of charge transport in intrinsic polycrystalline geometries. We created realistic models of large CVD-grown graphene samples and then computed the corresponding charge carrier mobilities as a function of the average grain size and the coalescence quality between the grains. Our results reveal a remarkably simple scaling law for the mean free path and conductivity, correlated to atomic-scale charge density fluctuations along grain boundaries.

Furthermore, we used numerical simulations and transport measurements to demonstrate that electrical properties and chemical modification of graphene grain boundaries are strongly correlated. This not only provides guidelines for the improvement of graphene devices, but also opens a new research area of engineering graphene grain boundaries for highly sensitive electro-biochemical devices.

We investigated the charge transport properties of planar amorphous graphene that is fully topologically disordered, in the form of sp^2 threefold coordinated networks consisting of hexagonal rings but also including many pentagons and heptagons

© Springer International Publishing Switzerland 2016
D.V. Tuan, *Charge and Spin Transport in Disordered Graphene-Based Materials*, Springer Theses, DOI 10.1007/978-3-319-25571-2_6

distributed in a random fashion. Using the Kubo transport methodology and the Lanczos method, the density of states, mean free paths, and semiclassical conductivities of such amorphous graphene membranes are computed. Despite a large increase in the density of states close to the charge neutrality point, all electronic properties are dramatically degraded, evidencing an Anderson insulating state caused by topological disorder alone. These results are supported by Landauer-Büttiker conductance calculations, which show a localization length as short as 5 nm.

We reported on the transition from a Quantum Spin Hall effect regime to a robust metallic state, upon segregation of thallium adatoms adsorbed onto a graphene surface and introducing giant enhancement of spin-orbit coupling. Our theoretical methodology combines efficient calculation of both the Landauer-Büttiker conductance and the Kubo-Greenwood conductivity, giving access to both edge and bulk transport physics in disordered thallium-functionalized graphene systems of realistic sizes. Our findings quantify the detrimental effects of adatoms clustering in observing the QSHE, but provide additional bulk signature of a robust metallic state with minimum bulk conductivity of about $4e^2/h$, which should be helpful for guiding further experiments.

Finally, we have developed a new spin-transport-simulation method to investigate the spin transport in graphene. We showed that the presence of a low density of randomly distributed adatoms (inducing local Rashba spin-orbit coupling) yields ultrafast spin relaxation times at the Dirac point, together with an unconventional relation between the spin and momentum relaxation times. Our quantum transport simulations showed that certain types of adatoms (such as Nickel or Gold impurities) trigger strong spin decoherence at the Dirac point, although the transport regime eventually reaches the ballistic limit. This phenomenon hitherto unknown is a new type of spin dephasing mechanism, driven by entanglement between spin and pseudospin degrees of freedom. Those findings bring an unprecedented insight of spin relaxation mechanisms in graphene, suggesting a possible origin of reported low spin relaxation times, and clarification of the controversial description of relaxation mechanisms in various types of graphene samples.

Appendix A
Time Evolution of the Wave Packet

This appendix presents how to calculate the evolution of the wave packet $\hat{U}(t)|\varphi_{RP}\rangle$ and $[\hat{X}, \hat{U}(t)]|\varphi_{RP}\rangle$ which are used in the application of the real space method to calculate the transport properties. In order to do that, we divide the time t into small time steps $T = t/N$ and approximate $\hat{U}(T)$ with the series of orthogonal Chebyshev polynomials $Q_n(\hat{H})$

$$\hat{U}(T) = e^{\frac{-i\hat{H}T}{\hbar}} = \sum_{n=0}^{\infty} c_n(T) Q_n(\hat{H}) \tag{A.1}$$

The original Chebyshev polynomials T_n which satisfy the recurrent relations

$$T_0(x) = 1 \tag{A.2}$$

$$T_1(x) = x \tag{A.3}$$

$$T_2(x) = 2x^2 - 1 \tag{A.4}$$

$$\vdots$$

$$T_{n+1}(x) = 2x\, T_n(x) - T_{n-1}(x) \tag{A.5}$$

and act on the interval $[-1; 1]$ are rescaled to the rescaled Chebyshev polynomials Q_n which cover the bandwidth of system Hamiltonian $E \in [a - 2b : a + 2b]$, with the band center and bandwidth are a and $4b$, respectively. These rescaled Chebyshev polynomials Q_n satisfy

$$Q_n(E) = \sqrt{2} T_n \left(\frac{E - a}{2b} \right) \quad (\forall n \geq 1) \tag{A.6}$$

$$Q_0(E) = 1 \tag{A.7}$$

$$Q_1(E) = \sqrt{2} \frac{E - a}{2b} \tag{A.8}$$

© Springer International Publishing Switzerland 2016
D.V. Tuan, *Charge and Spin Transport in Disordered Graphene-Based Materials*, Springer Theses, DOI 10.1007/978-3-319-25571-2

$$Q_2(E) = 2\sqrt{2}\left(\frac{E-a}{2b}\right)^2 - \sqrt{2} \tag{A.9}$$

$$\vdots$$

$$Q_{n+1}(E) = 2\left(\frac{E-a}{2b}\right)Q_n(E) - Q_{n-1}(E) \tag{A.10}$$

With above definition, we have the orthonormal relations for $Q_n(E)$

$$\int Q_n(E)Q_m(E)p_Q(E)dE = \delta_{mn} \tag{A.11}$$

with respect to the weight

$$p_Q(E) = \frac{1}{2\pi b\sqrt{1 - \left(\frac{E-a}{2b}\right)^2}} \tag{A.12}$$

Once the Q_n polynomials are well defined, one can compute the related $c_n(T)$ coefficients

$$c_n(T) = \int dE \; p_Q(E)Q_n(E)e^{-i\frac{E}{\hbar}T} \tag{A.13}$$

$$= \int dE \; \frac{\sqrt{2}T_n\left(\frac{E-a}{2b}\right)}{2\pi b\sqrt{1 - \left(\frac{E-a}{2b}\right)^2}}e^{-i\frac{E}{\hbar}T} \tag{A.14}$$

$$= \frac{\sqrt{2}}{\pi}\int_{-1}^{1} dx \; \frac{T_n(x)}{\sqrt{1-x^2}}e^{-i\frac{(2bx+a)}{\hbar}T} \tag{A.15}$$

$$= \sqrt{2}i^n e^{-i\frac{a}{\hbar}T}J_n\left(-\frac{2b}{\hbar}T\right), \quad n \geq 1 \tag{A.16}$$

and the first coefficients $c_0(T) = i^n e^{-i\frac{a}{\hbar}T}J_0\left(-\frac{2b}{\hbar}T\right)$ with $J_n(x)$ is the Bessel function of the first kind and order n

We can now calculate $|\varphi_{RP}(T)\rangle$

$$|\varphi_{RP}(T)\rangle = \hat{U}(T)|\varphi_{RP}\rangle \tag{A.17}$$

$$|\varphi_{RP}(T)\rangle \simeq \sum_{n=0}^{N} c_n(T)Q_n(\hat{H})|\varphi_{RP}\rangle = \sum_{n=0}^{N} c_n(T)|\alpha_n\rangle \tag{A.18}$$

where $|\alpha_n\rangle = Q_n(\hat{H})|\varphi_{RP}\rangle$. With the definitions introduced in Eqs. (A.7, A.8 and A.9) and the recurrence relation Eq. (A.10), we obtain

$$|\alpha_0\rangle = |\varphi_{RP}\rangle \qquad (A.19)$$

$$|\alpha_1\rangle = \left(\frac{\hat{H} - a}{\sqrt{2}b}\right)|\alpha_0\rangle \qquad (A.20)$$

$$|\alpha_2\rangle = \left(\frac{\hat{H} - a}{b}\right)|\alpha_1\rangle - \sqrt{2}|\alpha_0\rangle \qquad (A.21)$$

$$|\alpha_{n+1}\rangle = \left(\frac{\hat{H} - a}{b}\right)|\alpha_n\rangle - |\alpha_{n-1}\rangle \quad (\forall n \geq 2) \qquad (A.22)$$

Following the same reasoning as for $|\varphi_{RP}(T)\rangle$, $|\varphi'_{RP}(T)\rangle$ can be evaluated first writting

$$|\varphi'_{RP}(T)\rangle = [\hat{X}, \hat{U}(T)]|\varphi_{RP}\rangle \qquad (A.23)$$

$$|\varphi'_{RP}(T)\rangle \simeq \sum_{n=0}^{N} c_n(T)[\hat{X}, Q_n(\hat{H})]|\varphi_{RP}\rangle = \sum_{n=0}^{N} c_n(T)|\beta_n\rangle \qquad (A.24)$$

with $|\beta_n\rangle = [\hat{X}, Q_n(\hat{H})]|\varphi_{RP}\rangle$. Using the Eqs. (A.10) and (A.19–A.22), we obtain the recurrence relation for $|\beta_n\rangle$

$$|\beta_0\rangle = 0 \qquad (A.25)$$

$$|\beta_1\rangle = \frac{\left[\hat{X}, \hat{H}\right]}{\sqrt{2}b}|\varphi_{RP}\rangle \qquad (A.26)$$

$$|\beta_{n+1}\rangle = \left(\frac{\hat{H} - a}{b}\right)|\beta_n\rangle - |\beta_{n-1}\rangle + \frac{1}{b}[\hat{X}, \hat{H}]|\alpha_n\rangle \quad (\forall n \geq 1) \qquad (A.27)$$

which contain $|\alpha_n\rangle$ and the commutator $[\hat{X}, \hat{H}]$ determined by the hopings and the distances between neighbours

$$[\hat{X}, \hat{\mathcal{H}}] = \begin{pmatrix} 0 & & & \\ & \ddots & \mathcal{H}_{ij}\Delta X_{ij} & \\ & & \ddots & \\ \mathcal{H}_{ji}\Delta X_{ji} & & & \ddots \\ & & & 0 \end{pmatrix} \qquad (A.28)$$

where $\Delta X_{ij} = (X_i - X_j)$ is the distance between orbitals $|\varphi_i\rangle$ and $|\varphi_j\rangle$.

Appendix B
Lanczos Method

In this appendix the Lanczos method is introduced. Instead of diagonalizing the Hamiltonian the Lanczos method is a useful method to transform the Hamiltonian into tridiagonal matrix which is more convenient to compute the density of state or spin polarization. The general idea of this method is building from the initial state $|\varphi_{RP}\rangle$ a new basis in which the Hamiltonian is tridiagonal. Here are the basic steps:

The first step starts with the first vector in the new basis $|\psi_1\rangle = |\varphi_{RP}\rangle$ and builds the second one $|\psi_2\rangle$ which is orthonormal to the first one

$$a_1 = \langle \psi_1 | \hat{H} | \psi_1 \rangle \tag{B.1}$$

$$|\tilde{\psi}_2\rangle = \hat{H}|\psi_1\rangle - a_1|\psi_1\rangle \tag{B.2}$$

$$b_1 = \||\tilde{\psi}_2\rangle\| = \sqrt{\langle \tilde{\psi}_2 | \tilde{\psi}_2 \rangle} \tag{B.3}$$

$$|\psi_2\rangle = \frac{1}{b_1}|\tilde{\psi}_2\rangle \tag{B.4}$$

All other recursion steps ($\forall n \geq 1$) are identical, we build the $(n+1)^{th}$ vector which is orthonormal to the previous ones and given by

$$a_n = \langle \psi_n | \hat{H} | \psi_n \rangle \tag{B.5}$$

$$|\tilde{\psi}_{n+1}\rangle = \hat{H}|\psi_n\rangle - a_n|\psi_n\rangle - b_{n-1}|\psi_{n-1}\rangle \tag{B.6}$$

$$b_n = \sqrt{\langle \tilde{\psi}_{n+1} | \tilde{\psi}_{n+1} \rangle} \tag{B.7}$$

$$|\psi_{n+1}\rangle = \frac{1}{b_n}|\tilde{\psi}_{n+1}\rangle \tag{B.8}$$

The coefficients a_n and b_n are named recursion coefficients which are respectively the diagonal and off-diagonal of the matrix representation of $\hat{H}$ in the Lanczos basis (that we write $\tilde{\hat{H}}$).

© Springer International Publishing Switzerland 2016
D.V. Tuan, *Charge and Spin Transport in Disordered Graphene-Based Materials*, Springer Theses, DOI 10.1007/978-3-319-25571-2

$$\tilde{\hat{H}} = \begin{pmatrix} a_1 & b_1 & & & & \\ b_1 & a_2 & b_2 & & & \\ & b_2 & \ddots & \ddots & & \\ & & \ddots & \ddots & b_N & \\ & & & b_N & a_N & \end{pmatrix} \qquad (B.9)$$

With simple linear algebra, one shows that

$$\langle \varphi_{RP} | \delta(E - \hat{H}) | \varphi_{RP} \rangle = \langle \psi_1 | \delta(E - \hat{H}) | \psi_1 \rangle$$
$$= \lim_{\eta \to 0} -\frac{1}{\pi} \Im_m \left(\langle \psi_1 | \frac{1}{E + i\eta - \hat{H}} | \psi_1 \rangle \right)$$

while

$$\langle \psi_1 | \frac{1}{E + i\eta - \tilde{\hat{H}}} | \psi_1 \rangle = \cfrac{1}{E + i\eta - a_1 - \cfrac{b_1^2}{E + i\eta - a_2 - \cfrac{b_2^2}{E + i\eta - a_3 - \cfrac{b_3^2}{\ddots}}}}$$

$$(B.10)$$

which is referred as a continued fraction G_1 with the definition of G_n as,

$$G_n = \cfrac{1}{E + i\eta - a_n - \cfrac{b_n^2}{E + i\eta - a_{n+1} - \cfrac{b_{n+1}^2}{E + i\eta - a_{n+2} - \cfrac{b_{n+2}^2}{\ddots}}}} \qquad (B.11)$$

$$G_1 = \frac{1}{E + i\eta - a_1 - b_1^2 G_2} \qquad (B.12)$$

$$G_n = \frac{1}{E + i\eta - a_n - b_n^2 G_{n+1}} \qquad (B.13)$$

Since we compute a finite number of recursion coefficients, the subspace of Lanczos if of finite dimension (N), so it is crucial to terminate the continued fraction by an appropriate choice of the last $\{a_{n=N}, b_{n=N}\}$ elements. Let us rewrite the continued fraction as

$$G_1 = \cfrac{1}{E + i\eta - a_1 - \cfrac{b_1^2}{E + i\eta - a_2 - \cfrac{b_2^2}{E + i\eta - a_3 - \cfrac{b_3^2}{\ddots \cfrac{}{E + i\eta - a_N - b_N^2 G_{N+1}}}}}}$$

(B.14)

where G_{N+1} denotes such termination. The simplest case is when all the spectrum is contained in a finite bandwidth $[a - 2b; a + 2b]$, a the spectrum center and $4b$ its bandwidth. Recursion coefficients a_n and b_n oscillate around their average value a et b, and the damping is usually fast after a few hundreds of recursion steps. The termination then satisfies

$$G_{N+1} = \frac{1}{E + i\eta - a - b^2 G_{N+2}} = \frac{1}{E + i\eta - a - b^2 G_{N+1}} \tag{B.15}$$

from which a polynomial of second degree is found

$$-(b^2)G_{N+1}^2 + (E + i\eta - a)G_{N+1} - 1 = 0 \tag{B.16}$$

and straightforwardly solved

$$\Delta = (E + i\eta - a)^2 - (2b)^2 \tag{B.17}$$

$$G_{N+1} = \frac{(E + i\eta - a) \mp i\sqrt{-\Delta}}{2b^2} \tag{B.18}$$

$$G_{N+1} = \frac{(E + i\eta - a) - i\sqrt{(2b)^2 - (E + i\eta - a)^2}}{2b^2} \tag{B.19}$$

Curriculum Vitae

Dinh Van Tuan

Contact information

Postdoctoral Researcher   E-mail: tuan.dinh@icn.cat;
dinhvantuan1984@gmail.com
Theoretical and Computational Nanoscience Group,
Catalan Institute of Nanoscience and Nanotechnology

Research Interests

Quantum Condensed Matter Theory: Charge and spin transport, quantum Hall effect, spin Hall effect, topological electronic phases and disordered electronic systems.

Professional Preparation

- Ph.D., Materials Science, 2011–2014

  - Department of Physics, Autonomous University of Barcelona, Spain
  - Thesis Topic: *Charge and Spin Transport in Disordered Graphene-Based Materials*
  - Supervisor: Prof. Stephan Roche

© Springer International Publishing Switzerland 2016   151
D.V. Tuan, *Charge and Spin Transport in Disordered Graphene-Based
Materials*, Springer Theses, DOI 10.1007/978-3-319-25571-2

- M. Sc., Theoretical and Mathematical Physics, 2008-2010
  - Department of Theoretical Physics, Ho Chi Minh city University of Science, Vietnam
  - Thesis Topic: *The Graphene Polarizability and Applications*
  - Supervisor: Associate Prof. Nguyen Quoc Khanh

Professional Appointments

9/2014   Postdoctoral Researcher, Catalan Institute of Nanoscience and Nanotechnology
9/2011–9/2014 Ph.D. student, Catalan Institute of Nanoscience and Nanotechnology
9/2007–9/2011 Research Assistant, Ho Chi Minh city University of Science

Publications

1. **Dinh Van Tuan**, Frank Ortmann, David Soriano, Sergio O. Valenzuela, and Stephan Roche. Pseudospin-driven spin relaxation mechanism in graphene. *Nature Physics*, 10, 857–863 (2014)
2. Alessandro Cresti, David Soriano, **Dinh Van Tuan**, Aron W. Cummings, and Stephan Roche. Multiple Quantum Phases in Graphene with Enhanced Spin-Orbit Coupling: from Quantum Spin Hall Regime to Spin Hall Effect and Robust Metallic State. *Physical Review Letter*, 113, 246603 (2014)
3. Aron W. Cummings, Dinh Loc Duong, Van Luan Nguyen, **Dinh Van Tuan**, Jani Kotakoski, Jose Eduardo Barrios Vargas, Young Hee Lee and Stephan Roche. Charge Transport in Polycrystalline Graphene: Challenges and Opportunities. *Advanced Materials*, **26**, Issue 30, 5079–5094 (2014)
4. David Jiménez, Aron W. Cummings, Ferney Chaves, **Dinh Van Tuan**, Jani Kotakoski, and Stephan Roche. Impact of graphene polycrystallinity on the performance of graphene field-effect transistors. *Appl. Phys. Lett.*, **104**, 043509 (2014)
5. **Dinh Van Tuan**, Jani Kotakoski, Thibaud Louvet, Frank Ortmann, Jannik C. Meyer, and Stephan Roche. Scaling Properties of Charge Transport in Polycrystalline Graphene. *Nano Letters*, **13** (4), 1730–1735 (2013)
6. Alessandro Cresti, Frank Ortmann, Thibaud Louvet, **Dinh Van Tuan**, and Stephan Roche. Broken Symmetries, Zero-Energy Modes, and Quantum Transport in Disordered Graphene. *Phys. Rev. Lett.*, **110**, 196601 (2013).
7. Alessandro Cresti, Thibaud Louvet, Frank Ortmann, **Dinh Van Tuan**, Paweł Lenarczyk, Georg Huhs and Stephan Roche. Impact of Vacancies on Diffusive and Pseudodiffusive Electronic Transport in Graphene. *Crystals*, **3**, 289–305 (2013).
8. **Dinh Van Tuan** and Nguyen Quoc Khanh. Plasmon modes of double-layer graphene at finite temperature. *Physica E: Low-dimensional Systems and Nanostructures*, **54**, 267 (2013)

9. **Dinh Van Tuan**, Avishek Kumar, Stephan Roche, Frank Ortmann, M. F. Thorpe, and Pablo Ordejon. Insulating behavior of an amorphous graphene membrane. *Phys. Rev. B*, **86**, 121408 (Rapid Communications) (2012)

10. **Dinh Van Tuan** and Nguyen Quoc Khanh. Temperature effects on Plasmon modes of double-layer graphene. *Communications in Physics*, **22**, 45 (2012)

11. David Soriano, **Dinh Van Tuan**, Simon M.-M. Dubois, Martin Gmitra, Aron W. Cummings, Denis Kochan, Frank Ortmann, Jean-Christophe Charlier, Jaroslav Fabian, and Stephan Roche. Spin Transport in Hydrogenated Graphene. *accepted for publication in 2D Materials as a Topical Review*, (2015)

Honors and Awards

- Award for the best graduate student, 2010
- Vietnamese Ministry of Education award, 2005
- Sliver Medal at the National Physics Olympiad for the students of national universities, 2005
- Bronze Medal at the National Physics Olympiad, 2003

Printed in the United States
By Bookmasters